SCIENTIFIC ROMANCES.

BY

C. H. HINTON, B.A.

WHAT IS THE FOURTH DIMENSION?
THE PERSIAN KING.
A PLANE WORLD.
A PICTURE OF OUR UNIVERSE.
CASTING OUT THE SELF.

ARDVA·QVÆ·PVLCRA

Merchant Books
1886.

MB

ISBN 1-60386-157-2

What is the Fourth Dimension?

CHAPTER I.

AT the present time our actions are largely influenced by our theories. We have abandoned the simple and instinctive mode of life of the earlier civilisations for one regulated by the assumptions of our knowledge and supplemented by all the devices of intelligence. In such a state it is possible to conceive that a danger may arise, not only from a want of knowledge and practical skill, but even from the very presence and possession of them in any one department, if there is a lack of information in other departments. If, for instance, with our present knowledge of physical laws and mechanical skill, we were to build houses without regard to the conditions laid down by physiology, we should probably—to suit an apparent convenience—make them perfectly draught-tight, and the best-constructed mansions would be full of suffocating chambers. The knowledge of the construction of the body and the conditions of its health prevent it from suffering injury by the development of our powers over nature.

In no dissimilar way the mental balance is saved from the dangers attending an attention concentrated on the laws of mechanical science by a just consideration of the constitution of the knowing faculty, and the conditions of knowledge.   Whatever pursuit we are engaged in, we are acting consciously or unconsciously upon some theory, some view of things.   And when the limits of daily routine are continually narrowed by the ever-increasing complication of our civilisation, it becomes doubly important that not one only but every kind of thought should be shared in.

There are two ways of passing beyond the domain of practical certainty, and of looking into the vast range of possibility.   One is by asking, "What is knowledge? What constitutes experience?"   If we adopt this course we are plunged into a sea of speculation.   Were it not that the highest faculties of the mind find therein so ample a range, we should return to the solid ground of facts, with simply a feeling of relief at escaping from so great a confusion and contradictoriness.

The other path which leads us beyond the horizon of actual experience is that of questioning whatever seems arbitrary and irrationally limited in the domain of knowedge.   Such a questioning has often been successfully applied in the search for new facts.   For a long time four gases were considered incapable of being reduced to the liquid state.   It is but lately that a physicist has succeeded in showing that there is no such arbitrary distinction among gases.   Recently again the question has been raised, "Is there not a fourth state of matter?"   Solid, liquid, and gaseous states are known.   Mr. Crookes attempts to demonstrate the existence of a state differing from all of these.   It is the object of these pages to show that, by supposing away certain limitations of the fundamental conditions of existence as we know it, a state of being can be conceived with powers far transcending our own. When this is made clear it will not be out of place to n vestigate what relations would subsist between our mode of existence and that which will be seen to be a possible one.

In the first place, what is the limitation that we must suppose away ?

An observer standing in the corner of a room has three directions naturally marked out for him ; one is upwards along the line of meeting of the two walls ; another is forwards where the floor meets one of the walls ; a third is sideways where the floor meets the other wall. He can proceed to any part of the floor of the room by moving first the right distance along one wall, and then by turning at right angles and walking parallel to the other wall. He walks in this case first of all in the direction of one of the straight lines that meet in the corner of the floor, afterwards in the direction of the other. By going more or less in one direction or the other, he can reach any point on the floor, and any movement, however circuitous, can be resolved into simple movements in these two directions.

But by moving in these two directions he is unable to raise himself in the room. If he wished to touch a point in the ceiling, he would have to move in the direction of the line in which the two walls meet. There are three directions then, each at right angles to both the other, and entirely independent of one another. By moving in these three directions or combinations of them, it is possible to arrive at any point in a room. And if we suppose the straight lines which meet in the corner of the room to be prolonged indefinitely, it would be possible by moving in the direction of those three lines, to arrive at any point in space. Thus in space there are three independent directions, and only three ; every other direction is compounded of these three. The question that comes before us then is this. " Why

should there be three and only three directions ?"   Space,
as we know it, is subject to a limitation.

In order to obtain an adequate conception of what
this limitation is, it is necessary to first imagine beings
existing in a space more limited than that in which we
move.   Thus we may conceive a being who has been
throughout all the range of his experience confined to a
single straight line.   Such a being would know what it
was to move to and fro, but no more.   The whole of
space would be to him but the extension in both directions of the straight line to an infinite distance.   It is
evident that two such creatures could never pass one
another.   We can conceive their coming out of the
straight line and entering it again, but they having
moved always in one straight line, would have no conception of any other direction of motion by which such
a result could be effected.   The only shape which could
exist in a one-dimensional existence of this kind would
be a finite straight line.   There would be no difference in
the shapes of figures ; all that could exist would simply
be longer or shorter straight lines.

Again, to go a step higher in the domain of a conceivable existence.   Suppose a being confined to a plane
superficies, and throughout all the range of its experience
never to have moved up or down, but simply to have kept
to this one plane.   Suppose, that is, some figure, such
as a circle or rectangle, to be endowed with the power
of perception ; such a being if it moves in the plane
superficies in which it is drawn, will move in a multitude
of directions ; but, however varied they may seem to be,
these directions will all be compounded of two, at right
angles to each other.   By no movement so long as the
plane superficies remains perfectly horizontal, will this
being move in the direction we call up and down.   And
it is important to notice that the plane would be different,

to a creature confined to it, from what it is to us. We
think of a plane habitually as having an upper and a
lower side, because it is only by the contact of solids
that we realize a plane.   But a creature which had been
confined to a plane during its whole existence would
have no idea of there being two sides to the plane he
lived in.   In a plane there is simply length and breadth.
If a creature in it be supposed to know of an up or down
he must already have gone out of the plane.

Is it possible, then, that a creature so circumstanced
would arrive at the notion of there being an up and down,
a direction different from those to which he had been
accustomed, and having nothing in common with them ?
Obviously nothing in the creature's circumstances would
tell him of it.   It could only be by a process of reasoning
on his part that he could arrive at such a conception.   If
he were to imagine a being confined to a single straight
line, he might realise that he himself could move in two
directions, while the creature in a straight line could only
move in one.   Having made this reflection he might ask,
" But why is the number of directions limited to two ?
Why should there not be three ? "

A creature (if such existed), which moves in a plane
would be much more fortunately circumstanced than one
which can only move in a straight line.   For, in a plane,
there is a possibility of an infinite variety of shapes, and
the being we have supposed could come into contact
with an indefinite number of other beings.   He would
not be limited, as in the case of the creature in a straight
line, to one only on each side of him.

It is obvious that it would be possible to play curious
tricks with a being confined to a plane.   If, for instance,
we suppose such a being to be inside a square, the only
way out that he could conceive would be through one of
the sides of the square.   If the sides were impenetrable,

he would be a fast prisoner, and would have no way out.

What his case would be we may understand, if we reflect what a similar case would be in our own existence. The creature is shut in in all the directions he knows of. If a man is shut in in all the directions he knows of, he must be surrounded by four walls, a roof and a floor. A two-dimensional being inside a square would be exactly in the same predicament that a man would be, if he were in a room with no opening on any side. Now it would be possible to us to take up such a being from the inside of the square, and to set him down outside it. A being to whom this had happened would find himself outside the place he had been confined in, and he would not have passed through any of the boundaries by which he was shut in. The astonishment of such a being can only be imagined by comparing it to that which a man would feel, if he were suddenly to find himself outside a room in which he had been, without having passed through the window, doors, chimney or any opening in the walls, ceiling or floor.

Another curious thing that could be effected with a two-dimensional being, is the following. Conceive two beings at a great distance from one another on a plane surface. If the plane surface is bent so that they are brought close to one another, they would have no conception of their proximity, because to each the only possible movements would seem to be movements in the surface. The two beings might be conceived as so placed, by a proper bending of the plane, that they should be absolutely in juxtaposition, and yet to all the reasoning faculties of either of them a great distance could be proved to intervene. The bending might be carried so far as to make one being suddenly appear in the plane by the side of the other. If these beings were ignorant of the existence of a third dimension, this result would be as marvellous to them, as it would be for a human being who was at a great distance—it might be at the other side of the world—to suddenly appear and really be by our side, and during the whole time he not to have left the place in which he was.

CHAPTER II.

THE foregoing examples make it clear that beings can be conceived as living in a more limited space than ours. Is there a similar limitation in the space we know?

At the very threshold of arithmetic an indication of such a limitation meets us.

If there is a straight line before us two inches long, its length is expressed by the number 2. Suppose a square to be described on the line, the number of square inches in this figure is expressed by the number 4, *i.e.*, 2 × 2. This 2 × 2 is generally written 2^2, and named " 2 square."

Now, of course, the arithmetical process of multiplication is in no sense identical with that process by which a square is generated from the motion of a straight line, or a cube from the motion of a square. But it has been observed that the units resulting in each case, though different in kind, are the same in number.

If we touch two things twice over, the act of touching has been performed four times. Arithmetically, 2 × 2 = 4. If a square is generated by the motion of a line two inches in length, this square contains four square inches.

So it has come to pass that the second and third powers of numbers are called "square" and "cube."

We have now a straight line two inches long. On this a square has been constructed containing four square inches. If on the same line a cube be constructed, the number of cubic inches in the figure so made is 8, *i.e.*, 2 × 2 × 2 or 2^3. Here, corresponding to the numbers 2, 2^2, 2^3, we have a series of figures. Each figure contains more units than the last, and in each the unit is of a different kind. In the first figure a straight line is the unit, viz., one linear inch; it is said to be of one dimension. In the second a square is the unit, viz., one square inch. The square is a figure of two dimensions. In the third case a cube is the unit, and the cube is of three dimensions. The straight line is said to be of one dimension because it can be measured only in one way. Its length can be taken, but it has no breadth or thickness. The square is said to be of two dimensions because it has both length and breadth. The cube is said to have three dimensions, because it can be measured in three ways.

The question naturally occurs, looking at these numbers 2, 2^2, 2^3, by what figure shall we represent 2^4, or 2 × 2 × 2 × 2. We know that in the figure there must be sixteen units, or twice as many units as in the cube. But the unit also itself must be different. And it must not differ from a cube simply in shape. It must differ from a cube as a cube differs from a square. No number of squares will make up a cube, because each square has no thickness. In the same way, no number of cubes must be able to make up this new unit. And here, instead of trying to find something already known, to which the idea of a figure corresponding to the fourth power can be affixed, let us simply reason out what the properties of such a figure must be. In this attempt we have to rely, not on a process of touching or vision, such as informs us of the properties of bodies in the space we know, but on a process of thought. Each fact concerning this unknown figure has to be reasoned out; and it is only after a number of steps have been gone through, that any consistent familiarity with its properties is obtained. Of all applications of the reason, this exploration is perhaps the one which requires, for the simplicity of the data involved, the greatest exercise of the abstract imagination, and on this account is well worth patient attention. The first steps are very simple. We must imagine a finite straight line to generate a square by moving on the plane of the paper, and this square in its turn to generate a cube by moving vertically upwards. Fig. 1 represents a straight line; Fig. 2 represents a square formed by the motion of that straight line; Fig. 3 represents perspectively a cube

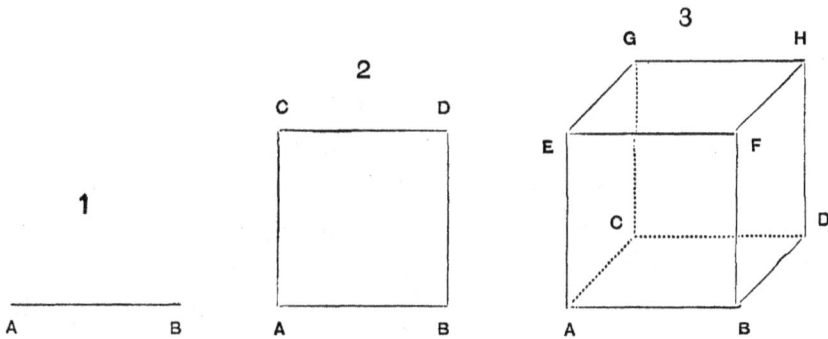

formed by the motion of that square A B C D upwards. It would be well, instead of using figure 3, to place a cube on the paper. Its base would be A B C D, its upper surface E F G H.

The straight line A B gives rise to the square A B C D by a movement at right angles to itself. If motion be confined to the straight line A B, a backward and forward motion is the only one possible. No sideway motion is admissible. And if we suppose a being to exist which could only move in the straight line A B, it would have no idea of any other movement than to and fro. The square A B C D is formed from the straight line by a movement in a direction entirely different from the direction which exists in A B.   This motion is not expressible by means of any possible motion in A B.   A being which existed in A B, and whose experience was limited to what could occur in A B, would not be able to understand the instructions we should give to make A B trace out the figure A B C D.

In the figure A B C D there is a possibility of moving in a variety of directions, so long as all these directions are confined to one plane.   All directions in this plane can be considered as compounded of two, from A to B, and from A to C.   Out of the infinite variety of such directions there is none which tends in a direction perpendicular to Fig. 2 ; there is none which tends upwards from the plane of the paper.   Conceive a being to exist in the plane, and to move only in it.   In all the movements which he went through there would be none by which he could conceive the alteration of Fig. 2 into what Fig. 3 represents in perspective.   For 2 to become 3 it must be supposed to move perpendicularly to its own plane.   The figure it traces out is the cube A B C D E F G H.

All the directions, manifold as they are, in which a creature existing in Fig. 3 could move, are compounded of three directions.   From A to B, from A to C, from A to E, and there are no other directions known to it.

But if we suppose something similar to be done to Fig. 3, something of the same kind as was done to Fig. 1 to turn it into Fig. 2, or to Fig. 2 to turn it into Fig. 3, we must suppose the whole figure as it exists to be moved in some direction entirely different from any direction within it, and not made up of any combination of the directions in it.   What is this ?   It is the fourth direction.

We are as unable to imagine it as a creature living in the plane Fig. 2 would be to imagine a direction such that moving in it the square 2 would become the cube 3. The third dimension to such a creature would be as unintelligible as the fourth is to us. And at this point we have to give up the aid that is to be got from any presentable object, and we have simply to investigate what the properties of the simplest figure in four dimensions are, by pursuing further the analogy which we know to exist between the process of formation of 2 from 1, and of 3 from 2, and finally of 4 from 3. For the sake of convenience, let us call the figure we are investigating—the simplest figure in four dimensions—a four-square.

First of all we must notice, that if a cube be formed from a square by the movement of the square in a new direction, each point of the interior of the square traces out part of the cube. It is not only the bounding lines that by their motion form the cube, but each portion of the interior of the square generates a portion of the cube. So if a cube were to move in the fourth dimension so as to generate a four-square, every point in the interior of the cube would start *de novo*, and trace out a portion of the new figure uninterfered with by the other points.

Or, to look at the matter in another light, a being in three dimensions, looking down on a square, sees each part of it extended before him, and can touch each part without having to pass through the surrounding parts, for he can go from above, while the surrounding parts surround the part he touches only in one plane.

So a being in four dimensions could look at and touch every point of a solid figure. No one part would hide another, for he would look at each part from a direction which is perfectly different from any in which it is possible to pass from one part of the body to another. To pass from one part of the body to another it is necessary to move in three directions, but a creature in four dimensions would look at the solid from a direction which is none of these three.

Let us obtain a few facts about the fourth figure,

proceeding according to the analogy that exists between
1, 2, 3, and 4.   In the Fig. 1 there are two points.   In 2
there are four points—the four corners of the square.   In
3 there are eight points.   In the next figure, proceeding
according to the same law, there would be sixteen points.

In the Fig. 1 there is one line.   In the square there
are four lines.   In the cube there are twelve lines.   How
many lines would there be in the four-square ?   That is
to say that there are three numbers—1, 4, and 12.   What
is the fourth, going on accordingly to the same law ?

To answer this question let us trace out in more detail
how the figures change into one another.   The line, to
become the square, moves ; it occupies first of all its
original position, and last of all its final position.   It
starts as A B, and ends as C D ; thus the line appears
twice, or it is doubled.   The two other lines in the square,
A C, B D, are formed by the motions of the points at
the extremities of the moving line.   Thus, in passing
from the straight line to the square the lines double
themselves, and each point traces out a line.   If the
same procedure holds good in the case of the change of
the square into the cube, we ought in the cube to have
double the number of lines as in the square—that is eight
—and every point in the square ought to become a line.
As there are four points in the square, we should have
four lines in the cube from them, that is, adding to the
previous eight, there should be twelve lines in the cube.
This is obviously the case.   Hence we may with confidence, to deduce the number of lines in a four-square,
apply this rule.   *Double the number of lines in the previous figure, and add as many lines as there are points
in the previous figure.*   Now in the cube there are twelve
lines and eight points.   Hence we get 2 × 12 + 8, or
thirty-two lines in the four-square.

In the same way any other question about the four-square can be answered. We must throw aside our realising power and answer in accordance with the analogy to be worked out from the three figures we know.

Thus, if we want to know how many plane surfaces the four-square has, we must commence with the line, which has none; the square has one; the cube has six. Here we get the three numbers, 0, 1, and 6. What is the fourth?

Consider how the planes of the cube arise. The square at the beginning of its motion determines one of the faces of the cube, at the end it is the opposite face, during the motion each of the lines of the square traces out one plane face of the cube. Thus we double the number of planes in the previous figure, and every line in the previous figure traces out a plane in the subsequent one.

Apply this rule to the formation of a square from a line. In the line there is no plane surface, and since twice nothing is nothing, we get, so far, no surface in the square; but in the straight line there is one line, namely itself, and this by its motion traces out the plane surface of the square. So in the square, as should be, the rule gives one surface.

Applying this rule to the case of the cube, we get, doubling the surfaces, 12; and adding a plane for each of the straight lines, of which there are 12, we have another 12, or 24 plane surfaces in all. Thus, just as by handling or looking at it, it is possible to describe a figure in space, so by going through a process of calculation it is within our power to describe all the properties of a figure in four dimensions.

There is another characteristic so remarkable as to need a special statement. In the case of a finite straight line, the boundaries are points. If we deal with one dimension only, the figure 1, that of a segment of a straight line, is cut out of and separated from the rest of an imaginary infinitely long straight line by the two points at its extremities. In this simple case the two points correspond to the bounding surface of the cube. In the case of a two-dimensional figure an infinite plane represents the whole of space. The square is separated off by four straight lines, and it is impossible for an entry to be made into the interior of the square, except by passing through the straight lines. Now, in these cases, it is evident that the boundaries of the figure are of one dimension less than the figure itself. Points bound lines, lines bound plane figures, planes bound solid figures. Solids then must bound four dimensional figures. The four-square will be bounded in the following manner. First of all there is the cube which, by its motion in the fourth direction, generates the figure. This, in its initial position, forms the base of the four-square. In its final position it forms the opposite end. During the motion each of the faces of the cube give rise to another cube. The direction in which the cube moves is such that of all the six sides none is in the least inclined in that direction. It is at right angles to all of them. The base of the cube, the top of the cube, and the four sides of the cube, each and all of them form cubes. Thus the four-square is bounded by eight cubes. Summing up, the four-square would have 16 points, 32 lines, 24 surfaces, and it would be bounded by 8 cubes.

If a four-square were to rest in space it would seem to us like a cube.

To justify this conclusion we have but to think of how a cube would appear to a two-dimensional being. To come within the scope of his faculties at all, it must come into contact with the plane in which he moves. If it is brought into as close a contact with this plane as possible, it rests on it by one of its faces. This face is a square, and the most a two-dimensional being could get acquainted with of a cube would be a square.

Having thus seen how it is possible to describe the properties of the simplest shape in four dimensions, it is evident that the mental construction of more elaborate figures is simply a matter of time and patience.

In the study of the form and development of the chick in the egg, it is impossible to detect the features that are sought to be observed, except by the use of the microscope. The specimens are accordingly hardened by a peculiar treatment and cut into thin sections. The investigator going over each of these sections, noticing all their peculiarities, constructs in his mind the shape as it originally existed from the record afforded by an indefinite number of slices. So, to form an idea of a four-dimensional figure, a series of solid shapes bounded on every side differing gradually from one another, proceeding, it may be, to the most diverse forms, has to be mentally grasped and fused into a unitary conception.

If, for instance, a small sphere were to appear, this to be replaced by a larger one, and so on, and then, when the largest had appeared, smaller and smaller ones to make their appearance, what would be witnessed would be a series of sections of a four-dimensional sphere. Each section in space being a sphere.

Again, just as solid figures can be represented on paper by perspective, four-dimensional figures can be represented perspectively by solids. If there are two squares, one lying over the other, and the underneath one be pushed away, its sides remaining parallel with the one that was over it, then if each point of the one be joined to the corresponding point of the other, we have a fair representation on paper of a cube. Fig. 3 may be considered to be such a representation if the square C D G H be considered to be the one that has been pushed away from lying originally under the square A B E F. Each of the planes which bound the cube is represented on the paper.

The only thing that is wanting is the three-dimensional content of the cube. So if two cubes be placed with their sides parallel, but one somewhat diagonally with regard to the other, and all their corresponding points be supposed joined, there will be found a set of solid figures, each representing (though of course distortedly) the bounding cubes of the four-dimensional figure, and every plane and line in the four-dimensional figure will be found to be represented in a kind of solid perspective. What is wanting is of course the four-dimensional content.

CHAPTER III.

HAVING now passed in review some of the properties of four-dimensional figures, it remains to ask what relations beings in four dimensions, if they did exist, would have with us.

And in the first place, a being in four dimensions would have to us exactly the appearance of a being in space. A being in a plane would only know solid objects as two-dimensional figures—the shapes namely in which they intersected his plane. So if there were four-dimensional objects, we should only know them as solids—the solids, namely, in which they intersect our space. Why, then, should not the four-dimensional beings be ourselves, and our successive states the passing of them through the three-dimensional space to which our consciousness is confined?

Let us consider the question in more detail. And for the sake of simplicity transfer the problem to the case of three and two dimensions instead of four and three.

Suppose a thread to be passed through a thin sheet of wax placed horizontally. It can be passed through in two ways. Either it can be pulled through, or it can be held at both ends, and moved downwards as a whole. Suppose a thread to be grasped at both ends, and the hands to be moved downwards perpendicularly to the sheet of wax. If the thread happens to be perpendicular to the sheet it simply passes through it, but if the thread be held, stretched slantingwise to the sheet, and the hands are moved perpendicularly downwards, the thread will, if it be strong enough, make a slit in the sheet.

If now the sheet of wax were to have the faculty of closing up behind the thread, what would appear in the sheet would be a moving hole.

Suppose that instead of a sheet and a thread, there were a straight line and a plane. If the straight line were placed slantingwise in reference to the plane and moved downwards, it would always cut the plane in a point, but that point of section would move on. If the plane were of such a nature as to close up behind the line, if it were of the nature of a fluid, what would be observed would be a moving point. If now there were a whole system of lines sloping in different directions, but all connected together, and held absolutely still by one framework, and if this framework with its system of lines were as a whole to pass slowly through the fluid plane at right angles to it, there would then be the appearance of a multitude of moving points in the plane, equal in number to the number of straight lines in the system. The lines in the framework will all be moving at the same rate—namely, at the rate of the framework in which they are fixed. But the points in the plane will have different velocities. They will move slower or faster, according as the lines which give rise to them are more or less inclined to the plane. A straight line perpendicular to the plane will, on passing through, give rise to a stationary point. A straight line that slopes very much inclined to the plane will give rise to a point moving with great swiftness. The motions and paths of the points would be determined by the arrangement of the lines in the system. It is obvious that if two straight lines were placed lying across one another like the letter X, and if this figure were to be stood upright and passed through the plane, what would appear would be at first two points. These two points would approach one another. When the part where the two strokes of the X meet came into the plane, the two points would become one. As the upper part of the figure passed through, the two points would recede from one another.

If the line be supposed to be affixed to all parts of the

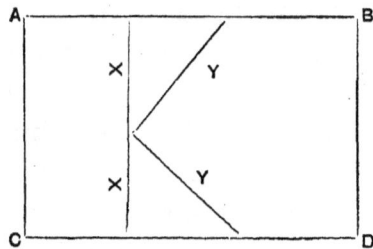

framework, and to loop over one another, and support one another,* it is obvious that they could assume all sorts of figures, and that the points on the plane would move in very complicated paths. The annexed figure represents a section of such a framework. Two lines X X and Y Y are shown, but there must be supposed to be a great number of others sloping backwards and forwards as well as sideways.

Let us now assume that instead of lines, very thin threads were attached to the framework: they on passing through the fluid plane would give rise to very small spots. Let us call the spots atoms, and regard them as

* A B C D framework, X and Y two lines interlinked

constituting a material system in the plane. There are four conditions which must be satisfied by these spots if they are to be admitted as forming a material system such as ours. For the ultimate properties of matter (if we eliminate attractive and repulsive forces, which may be caused by the motions of the smallest particles), are—1, Permanence; 2, Impenetrability; 3, Inertia; 4, Conservation of energy.

According to the first condition, or that of permanence, no one of these spots must suddenly cease to exist. That is, the thread which by sharing in the general motion of the system gives rise to the moving point, must not break off before the rest of them. If all the lines suddenly ended this would correspond to a ceasing of matter.

2. Impenetrability.—One spot must not pass through another. This condition is obviously satisfied. If the threads do not coincide at any point, the moving spots they give rise to cannot.

3. Inertia.—A spot must not cease to move or cease to remain at rest without coming into collision with another point. This condition gives the obvious condition with regard to the threads, that they, between the points where they come into contact with one another, must be straight. A thread which was curved would, passing through the plane, give rise to a point which altered in velocity spontaneously. This the particles of matter never do.

4. Conservation of energy.—The energy of a material system is never lost; it is only transferred from one form to another, however it may seem to cease. If we suppose each of the moving spots on the plane to be the unit of mass, the principle of the conservation of energy demands that when any two meet, the sum of the squares of their several velocities before meeting shall be the same as the

sum of the squares of their velocities after meeting. Now we have seen that any statement about the velocities of the spots in the plane is really a statement about the inclinations of the threads to the plane. Thus the principle of the conservation of energy gives a condition which must be satisfied by the inclinations of the threads of the plane. Translating this statement, we get in mathematical language the assertion that the sum of the squares of the tangents of the angles the threads make with the normal to the plane remains constant.

Hence, all complexities and changes of a material system made up of similar atoms in a plane could result from the uniform motion as a whole of a system of threads.

We can imagine these threads as weaving together to form connected shapes, each complete in itself, and these shapes as they pass through the fluid plane give rise to a series of moving points. Yet, inasmuch as the threads are supposed to form consistent shapes, the motion of the points would not be wholly random, but numbers of them would present the semblance of moving figures. Suppose, for instance, a number of threads to be so grouped as to form a cylinder for some distance, but after a while to be pulled apart by other threads with which they interlink. While the cylinder was passing through the plane, we should have in the plane a number of points in a circle. When the part where the threads deviated came to the plane, the circle would break up by the points moving away. These moving figures in the plane are but the traces of the shapes of threads as those shapes pass on. These moving figures may be conceived to have a life and a consciousness of their own.

Or, if it be irrational to suppose them to have a consciousness when the shapes of which they are momentary traces have none, we may well suppose that the shapes of threads have consciousness, and that the moving figures share this consciousness, only that in their case it is limited to those parts of the shapes that simultaneously pass through the plane. In the plane, then, we may conceive bodies with all the properties of a material system, moving and changing, possessing consciousness. After a while it may well be that one of them becomes so disassociated that it appears no longer as a unit, and its consciousness as such may be lost. But the threads of existence of such a figure are not broken, nor is the shape which gave it origin altered in any way. It has simply passed on to a distance from the plane. Thus nothing which existed in the conscious life on the plane would cease. There would in such an existence be no cause and effect, but simply the gradual realisation in a superficies of an already existent whole. There would be no progress, unless we were to suppose the threads as they pass to interweave themselves in more complex shapes.

Can a representation, such as the preceding, be applied to the case of the existence in space with which we have to do? Is it possible to suppose that the movements and changes of material objects are the intersections with a three-dimensional space of a four-dimensional existence? Can our consciousness be supposed to deal with a spatial profile of some higher actuality?

It is needless to say that all the considerations that have been brought forward in regard to the possibility of the production of a system satisfying the conditions of materiality by the passing of threads through a fluid plane, holds good with regard to a four-dimensional existence passing through a three-dimensional space. Each part of the ampler existence which passed through our space would seem perfectly limited to us. We should have no indication of the permanence of its existence.

Were such a thought adopted, we should have to imagine some stupendous whole, wherein all that has ever come into being or will come co-exists, which passing slowly on, leaves in this flickering consciousness of ours, limited to a narrow space and a single moment, a tumultuous record of changes and vicissitudes that are but to us. Change and movement seem as if they were all that existed. But the appearance of them would be due merely to the momentary passing through our consciousness of ever existing realities.

In thinking of these matters it is hard to divest ourselves of the habit of visual or tangible illustration. If we think of a man as existing in four dimensions, it is hard to prevent ourselves from conceiving him as prolonged in an already known dimension. The image we form resembles somewhat those solemn Egyptian statues which in front represent well enough some dignified sitting figure, but which are immersed to their ears in a smooth mass of stone which fits their contour exactly.

No material image will serve. Organised beings seem to us so complete that any addition to them would deface their beauty. Yet were we creatures confined to a plane, the outline of a Corinthian column would probably seem to be of a beauty unimprovable in its kind. We should be unable to conceive any addition to it, simply for the reason that any addition we could conceive would be of the nature of affixing an unsightly extension to some part of the contour. Yet, moving as we do in space of three dimensions, we see that the beauty of the stately column far surpasses that of any single outline. So all that we can do is to deny our faculty of judging of the ideal completeness of shapes in four dimensions.

CHAPTER IV.

LET us now leave this supposition of framework and threads. Let us investigate the conception of a four-dimensional existence in a simpler and more natural manner—in the same way that a two-dimensional being should think about us, not as infinite in the third dimension, but limited in three dimensions as he is in two. A being existing in four dimensions must then be thought to be as completely bounded in all four directions as we are in three. All that we can say in regard to the possibility of such beings is, that we have no experience of motion in four directions. The powers of such beings and their experience would be ampler, but there would be no fundamental difference in the laws of force and motion.

Such a being would be able to make but a part of himself visible to us, for a cube would be apprehended by a two-dimensional being as the square in which it stood. Thus a four-dimensional being would suddenly appear as a complete and finite body, and as suddenly disappear, leaving no trace of himself, in space, in the same way that anything lying on a flat surface, would, on being lifted, suddenly vanish out of the cognisance of beings, whose consciousness was confined to the plane. The object would not vanish by moving in any direction, but disappear instantly as a whole. There would be no barrier no confinement of our devising that would not be perfectly open to him. He would come and go at pleasure ; he would be able to perform feats of the most surprising kind. It would be possible by an infinite plane extending in all directions to divide our space into two portions absolutely separated from one another ; but a four-dimensional being would slip round this plane with the greatest ease.

To see this clearly, let us first take the analogous case

in three dimensions. Suppose a piece of paper to represent
a plane.   If it is infinitely extended in every direction, it
will represent an infinite plane.   It can be divided into
two parts by an infinite straight line.   A being confined
to this plane could not get from one part of it to the other
without passing through the line.   But suppose another
piece of paper laid on the first and extended infinitely, it
will represent another infinite plane.   If the being moves
from the first plane by a motion in the third dimension,
it will move into this new plane.   And in it it finds no
line.   Let it move to such a position that when it goes
back to the first plane it will be on the other side of the
line.   Then let it go back to the first plane.   It has
appeared now on the other side of the line which divides
the infinite plane into two parts.

Take now the case of four dimensions.   Instead of
bringing before the mind a sheet of paper conceive a solid
of three dimensions.   If this solid were to become infinite
it would fill up the whole of three-dimensional space.
But it would not fill up the whole of four-dimensional
space.   It would be to four-dimensional space what an
infinite plane is to three-dimensional space.   There could
be in four-dimensional space an infinite number of such
solids, just as in three-dimensional space there could be
an infinite number of infinite planes.

Thus, lying alongside our space, there can be conceived
a space also infinite in all three directions.   To pass from
one to the other a movement has to be made in the fourth
dimension, just as to pass from one infinite plane to
another a motion has to be made in the third dimension.

Conceive, then, corresponding to the first sheet of paper
mentioned above, a solid, and as the sheet of paper was
supposed to be infinitely extended in two dimensions,
suppose the solid to be infinitely extended in its three
dimensions, so that it fills the whole of space as we know it.

Now divide this infinite solid in two parts by an infinite plane, as the infinite plane of paper was divided in two parts by an infinite line. A being cannot pass from one part of this infinite solid to another, on the other side of this infinite plane, without going through the infinite plane, *so long as he keeps within the infinite solid.*

But suppose beside this infinite solid a second infinite solid, lying next to it in the fourth dimension, as the second infinite plane of paper was next to the first infinite plane in the third dimension. Let now the being that wants to get on the other side of the dividing plane move off in the fourth dimension, and enter the second infinite solid. In this second solid there is no dividing plane. Let him now move, so that coming back to the first infinite solid he shall be on the other side of the infinite plane that divides it into two portions. If this is done, he will now be on the other side of the infinite plane, without having gone through it.

In a similar way a being, able to move in four dimensions, could get out of a closed box without going through the sides, for he could move off in the fourth dimension, and then move about, so that when he came back he would be outside the box.

Is there anything in the world as we know it, which would indicate the possibility of there being an existence in four dimensions? No definite answer can be returned to this question. But it may be of some interest to point out that there are certain facts which might be read by the light of the fourth dimensional theory.

To make this clear, let us suppose that space is really four dimensional, and that the three-dimensional space we know is, in this ampler space, like a surface is in our space.

We should then be in this ampler space like beings confined to the surface of a plane would be in ours. Let us suppose that just as in our space there are centres of attraction whose influence radiates out in every direction, so in this ampler space there are centres of attraction whose influence radiates out in every direction.   Is there anything to be observed in nature which would correspond to the effect of a centre of attraction lying out of our space, and acting on all the matter in it?   The effect of such a centre of attraction would not be to produce motion in any known direction, because it does not lie off in any known direction.

Let us pass to the corresponding case in three and two dimensions, instead of four and three.   Let us imagine a plane lying horizontally, and in it some creatures whose experience was confined to it.   If now some water or other liquid were poured on to the plane, the creatures, becoming aware of its presence, would find that it had a tendency to spread out all over the plane.   In fact it would not be to them as a liquid is to us—it would rather correspond to a gas.   For a gas, as we know it, tends to expand in every direction, and gradually increase so as to fill the whole of space.   It exercises a pressure on the walls of any vessel in which we confine it.

The liquid on the plane expands in all the dimensions which the two-dimensional creatures on the plane know, and at the same time becomes smaller in the third dimension, its absolute quantity remaining unchanged.   In like manner we might suppose that gases (which by expansion become larger in the dimensions that we know) become smaller in the fourth dimension.

The cause in this case would have to be sought for in an attractive force, acting with regard to our space as the force of gravity acts with regard to a horizontal plane.

Can we suppose that there is a centre of attraction somewhere off in the fourth dimension, and that the gases, which we know are simply more mobile liquids, expanding out in every direction under its influence. This view receives a certain amount of support from the fact proved experimentally that there is no absolute line of demarcation between a liquid and a gas. The one can be made to pass into the other with no moment intervening in which it can be said that now a change of state has taken place.

We might then suppose that the matter we know extending in three dimensions has also a small thickness in the fourth dimension ; that solids are rigid in the fourth as in the other three dimensions ; that liquids are too coherent to admit of their spreading out in space, and becoming thinner in the fourth dimension, under the influence of an attractive centre lying outside of our space ; but that gases, owing to the greater mobility of their particles, are subject to its action, and spread out in space under its influence, in the same manner that liquids, under the influence of gravity, spread out on a plane.

Then the density of a gas would be a measure of the relative thickness of it in the fourth dimension : and the diminution of the density would correspond to a diminution of the thickness in the fourth dimension. Could this supposition be tested in any way ?

Suppose a being confined to a plane ; if the plane is moved far off from the centre of attraction lying outside it, he would find that liquids had less tendency to spread out than before.

Or suppose he moves to a distant part of the plane so that the line from his position to the centre of attraction lies obliquely to the plane ; he would find that in this position a liquid would show a tendency to spread out more in one direction than another.

Now our space considered as lying in four-dimensional space, as a plane does in three-dimensional space, may be shifted. And the expansive force of gases might be

found to be different at different ages.   Or, shifting as we
do our position in space during the course of the earth's
path round the sun, there might arise a sufficient difference
in our position in space, with regard to the attractive centre,
to make the expansive force of gases different at different
times of the year, or to cause them to manifest a greater
expansive force in one direction than in another.

But although this supposition might be worked out at
some length, it is hard to suppose that it could afford
any definite test of the physical existence of a fourth
dimension.   No test has been discovered which is decisive.
And, indeed, before searching for tests, a theoretical point
of the utmost importance has to be settled.   In discussing
the geometrical properties of straight lines and planes, we
suppose them to be respectively of one and two dimensions, and by so doing deny them any real existence.
A plane and a line are mere abstractions.   Every portion
of matter is of three dimensions.   If we consider beings
on a plane not as mere idealities, we must suppose them
to be of some thickness.   If their experience is to be
limited to a plane this thickness must be very small
compared to their other dimensions.   Transferring our
reasoning to the case of four dimensions, we come to a
curious result.

If a fourth dimension exists there are two possible
alternatives.

One is, that there being four dimensions, we have a
three-dimensional existence only.   The other is that we
really have a four-dimensional existence, but are not
conscious of it.   If we are in three dimensions only,
while there are really four dimensions, then we must be
relatively to those beings who exist in four dimensions, as
lines and planes are in relation to us.   That is, we must
be mere abstractions.   In this case we must exist only in
the mind of the being that conceives us, and our experience

must be merely the thoughts of his mind—a result which has apparently been arrived at, on independent grounds, by an idealist philosopher.

The other alternative is that we have a four-dimensional existence. In this case our proportions in it must be infinitely minute, or we should be conscious of them. If such be the case, it would probably be in the ultimate particles of matter, that we should discover the fourth dimension, for in the ultimate particles the sizes in the three dimensions are very minute, and the magnitudes in all four dimensions would be comparable.

The preceding two alternative suppositions are based on the hypothesis of the reality of four-dimensional existence, and must be conceived to hold good only on that hypothesis.

It is somewhat curious to notice that we can thus conceive of an existence relative to which that which we enjoy must exist as a mere abstraction.

Apart from the interest of speculations of this kind they have considerable value ; for they enable us to express in intelligible terms things of which we can form no image. They supply us, as it were, with scaffolding, which the mind can make use of in building up its conceptions. And the additional gain to our power of representation is very great.

Many philosophical ideas and doctrines are almost unintelligible because there is no physical illustration which will serve to express them. In the imaginary physical existence which we have traced out, much that philosophers have written finds adequate representation. Much of Spinoza's Ethics, for example, could be symbolized from the preceding pages.

Thus we may discuss and draw perfectly legitimate conclusions with regard to unimaginable things.

It is, of course, evident that these speculations present

no point of direct contact with fact.   But this is no reason why they should be abandoned.   The course of knowledge is like the flow of some mighty river, which, passing through the rich lowlands, gathers into itself the contributions from every valley.   Such a river may well be joined by a mountain stream, which, passing with difficulty along the barren highlands, flings itself into the greater river down some precipitous descent, exhibiting at the moment of its union the spectacle of the utmost beauty of which the river system is capable.   And such a stream is no inapt symbol of a line of mathematical thought, which, passing through difficult and abstract regions, sacrifices for the sake of its crystalline clearness the richness that comes to the more concrete studies.   Such a course may end fruitlessly, for it may never join the main course of observation and experiment.   But, if it gains its way to the great stream of knowledge, it affords at the moment of its union the spectacle of the greatest intellectual beauty, and adds somewhat of force and mysterious capability to the onward current.

PART I.

The Persian King.

CHAPTER I.

IN Persia there was once a king. On one occasion when he was out hunting he came to the narrow entrance of a valley. It was shut in on either side by vast hills, seemingly the spurs from the distant mountains. These great spurs spread out including a wide tract of land. Towards the entrance where he stood they approached one another, and ended in abrupt cliffs. Across the mouth of the valley stretched a deep ravine. The king, followed by courtiers, galloped along, searching a spot where the deep fissure might be shallower, so that descending into it he might reach the valley by ascending on the opposite side.

But at every point the ravine stretched downwards dark and deep, from cliff to cliff, shutting off all access to the valley.

At one point only was there a means of crossing. There were two masses of rocks, jutting out one from either side like the abutments of a natural bridge, and they seemed to meet in mid air.

The mass trembled and shook as the king spurred his horse over it, and the dislodged stones reverberated

3

from side to side of the chasm till the noise of their falling was lost.

Before the first of his courtiers could follow him one of the great piers or abutments gave way—the whole mass fell crashing down. The king was alone in the valley.

"So ho," he cried, "the kingdom of Persia is shrunk to this narrow spot!" and without troubling himself for the moment how he should return, he sped onward.

But when he had ridden far into the valley on his steed that could outnumber ten leagues in an hour, and had returned to the entrance of it, he saw no trace of a living soul on the opposite brink of the cleft. No sign was left, save a few reeds bent down by the passage of the mounted train, that any human being had stood on the opposite side for ages.

The evening came on apace. Yet no one returned. Again he rode far into the valley. For the most part it was covered with long grass, but here and there a thick and tangled mass of vegetation attested to a great luxuriance of soil, while the surface was intersected here and there with rivulets of clear water, which finally lost their way in the dark gorge over which he had just so rashly adventured. But on no side did the steep cliffs offer any promise of escape.

When the night came on he stretched himself beneath one of the few trees not far from the ravine, while his faithful horse stood tranquilly at his head.

He did not awake till the moon had risen. But then suddenly he started to his feet, and walking to the edge of the cleft, peered over to the land from whence he had come. For he thought he heard sounds of some kind that were not the natural ones of the rustling wind or the falling water. Looking out he saw clearly opposite to him an old man in ragged clothing, leaning against a rock, holding a long pipe in his hands, on which he now and again played a few wild notes.

"Oho, peasant!" cried the king. "Run and tell the head man of your village that the king bids him come directly, and will have him bring with him the longest ropes and the strongest throwers under him."

But the old man did not seem to give heed. Then the king cried, "Hearken, old man, run quickly and tell your master that the king is confined here, and will reward him beyond his dreams if he deliver him quickly."

Then the old man rose, and coming nearer to the edge of the ravine stood opposite, still playing at intervals some notes on his long pipe. And the king cried, "Canst thou hear? Dost thou dare to refuse to carry my commands? For I am the king of Persia. Who art thou?"

Then the old man made answer, putting his pipe aside: "I am he who appears only when a man has passed for ever beyond the ken of all that have known him. I am Demiourgos, the maker of men."

Then the king cried, "Mock me not, but obey my commands."

The old man made answer, "I do not mock thee; and oh, my Lord, thou hast moved the puppets I have made, and driven them so to dance on the surface of the earth that I would willingly obey thee. But it is not permitted me to pass between thee and the world of men thou hast known."

Then the king was silent.

At length he said, "If thou art really what thou sayest, show me what thou canst do; build me a palace."

The old man lifted his pipe in both his trembling hands, and began to blow.

It was a strange instrument, for it not only produced

the shriller sounds of the lute, and the piercing notes of the trumpet, but resounded with the hollow booming of great organ pipes, and amongst all came ever and again a sharp and sonorous clang as of some metal instrument resounding when it was struck.

And then the king was as one who enjoys the delights of thought.   For in thought, delicate shades, impalpable nuances are ever passing.   It is as the blended strains of an invisible orchestra, but more subtle far, that come and go in unexpected metres, and overwhelm you with their beauty when all seemed silent.   And lo, as the strains sound, outside—palpable, large as the firmament, or real as the smallest thing you can take up and know it is there—outside stands some existence revealed—to be known and returned to for ever.

So the king, listening to this music, felt that something was rising behind him.   And turning, beheld course after course of a great building.   Almost as soon as he had looked it had risen completed, finished to the last embossure on the windows, the tracery on the highest pinnacles.   All had happened while the old man was blowing on his pipe, and when he ceased all was perfect.

And yet the appearance was very strange, for a finished and seemingly habitable building rose out of waste unreclaimed soil, strewn with rocks and barren. No dwellings were near the palace to wait on it, no roads led to it or away from it.

"There should be houses around it, and roadways," said the king ; "make them, and fields sown with corn, and all that is necessary for a state."

Blowing on his pipe in regular recurrent cadences, the old man called up houses close together, than scattered singly along roads which stretched away into the distance, to be seen every here and there perfectly clearly where they ascended a rising ground.   And near at

hand could be distinguished fields of grain and pasture
land.

Yet as the king turned to walk towards the new scene,
the old man laughed. "All this is a dream," he cried;
"so much I can do, but not at once." And breathing
peals of music from his pipe, he said, "This can be, but
is not yet."

"What," asked the king, "is all a delusion?" and as he
asked everything sank down. There was no palace, no
houses or fields, only the steep precipice-locked valley,
whither the king had ridden; and his horse cowering
behind him.

Then the king cried, "Thou art some moonstruck
hermit, leading out a life of folly alone. Get thee to
the village thou knowest, and bring me help."

But the old man answered him saying, "Great king, I
am bound to obey thee, and all the creative might of
my being I lay at thy feet; and lo, in the midst of this
valley I make for thee beings such as I can produce.
And all that thou hast seen is as nothing to what I can
do for thee. The depths of the starry heavens have no
limit, nor what I do for thee. Hast thou ever in thy
life looked into the deep still ocean, and lost thy sight
in the unseen depths? Even so thou wilt find no end
in what I will give thee. Hast thou ever in thy life
sought the depths of thy love's blue eyes, and found
therein a world which stretched on endlessly? Even so
I bring all to thy feet. Now that all the gladness of the
world has departed from thee, behold, I am a more
willing servant than ever thou hast had."

And again he played, and a hut rose up with a patch
of cleared soil around it, and a spring near by.

Then the king said, "Here will I dwell, and if I am
to be cut off from the rest of the world, I will lead a
peaceful life in this valley."

The sun was rising, the sounds had ceased, and the old man had disappeared.

CHAPTER II.

HE made his way slowly to the patch of cultivated ground, he knocked at the door of the hut, and then he called out. No answer was made to the sound of his voice, he entered, and saw a rude, plain interior. There were two forms half lying, half propped up by the walls, and some domestic implements lay about. But when he spoke to the beings they did not answer, and when he touched their arms they fell powerless on the ground and remained there. A terrible fear came on the king lest he should become such as these. He left them and again sought a possible outlet, but fruitlessly. And that evening he sought the old man again and inquired what sort of beings these were.

"For though in form and body like children outwardly," said the king, "they do nothing and seem unable to move ; are they in an enchanted slumber ? "

Then the old man came near to the edge of the ravine and, speaking solemnly and low, said :

"O king, thou dost not yet know the nature of the place wherein thou art. For these children are like the children thou hast known always both in form and body. I have worked on them as far as is within my power. But here in this valley a law reigns which binds them in sleepfulness and powerlessness. For here in everything that is done there is as much pain as pleasure. If it is pleasant to tread a downward slope there is as much pain in ascending the upward slope. And in every action there is a pleasant part and a painful part, and in the tasting of every herb the beings feel a bitter taste and a sweet taste, so indistinguishably united that the pleasure and the pain of eating it are equally balanced. And as hunger increases the sense of the bitterness in the taste increases, so it is never more pleasant to eat than not to eat. Everything that can be done here affords no more pleasure than it does pain, from the greatest action down to the least movement. And the beings as I can make them, they follow pleasure and avoid pain. And if the pleasure and the pain are equal they do not move one way or the other."

"This is impossible," said the king.

"Nay," said the old man, "that it is as I have said I will prove to thee." And he explained to the king how it would be possible to stimulate the children to activity, for he showed him how he could divest anything that was done of part of its pain and render it more pleasurable than painful. "In this way thou canst lead the beings I have given thee to do anything," said the old man, "but the condition is that thou must take the painful part that thou sparest them thyself." And he bade the king cut himself of the reeds that grew by the side of the ravine, and told him that putting them between himself and any being would enable him to take a part of the pain and leave in their feeling the whole of the pleasure and the pain diminished by that part which he bore himself.

Then the king cut of the reeds that grew by the side of the ravine. He went to the hut where these beings lay, and, taking the reeds in his hand, he placed one between the child's frame and himself. And the child rose up and walked, while he himself felt a pain in his limbs. And he found that by taking a pain in each part of him the child would exercise that part ; if he wished the child to look at anything he, by bearing a pain in his eyes, made looking at it pleasurable to the child, and accordingly the child did look at the object he wished him to regard. And again, by bearing a bitter taste in his mouth he made the child feel eating as pleasant, and the child gathered fruits and ate them.

Then the king by using two reeds made both the children move, and they went together wheresoever he wished them. But they had not the slightest idea of the king's action on them. They recognized each other, and played with each other. They saw the king and had a certain regard for him, but of his action on them they knew nothing. For they felt his bearing the pain as this thing or that being pleasurable. They felt his action as a motive in themselves.

And all day long the king went with them, leading them through the valley, bearing the pain of each step, so that the children felt nothing but pleasure. But at nightfall he led them back to the rude dwelling where he had found them. He led them by taking the pain from their steps in that direction, and not taking any of the pain from steps in any other direction.

And when they had entered the dwelling-place he removed his reeds from them. Immediately they sank down into the state of apathy in which he had found them. They did not move.

And the king at nightfall sought again the side of the ravine.

Gazing across it he saw the sandy waste of the land from which he had come, he saw the great stones which were scattered about, looking pale and grey in the moonlight. And presently in the shadow of a rock near the opposite brink he discerned the form of the old man.

And he cried out to him, and bade him come near.

And when the old man stood opposite to him, he besought him to tell him how he could make the beings go through their movements of life without his bearing so much pain.

And the old man took his staff in his hand, and he held it out towards the king, over the depth.

"Behold, O king, thy secret," he cried. And with his other hand he smote the staff which was pointing down into the depths. The staff swung to and fro many times, and at last it came to rest again.

Then the king besought him to explain what this might signify.

"Thou hast been," replied the old man, "as one who, wishing to make a staff swing to and fro, has made every movement separately, raising it up by his hand each time that it falls down. But, behold, when I set it in movement it goes through many swings of itself, both downward and upward, until the movement I imparted to it is lost. Even so thou must make these beings go through both pleasure and pain, thyself bearing but the difference, not taking all the pain."

"Must I then," asked the king, "by bearing pain give these beings a certain store of pleasure, and then let them go through their various actions until they have exhausted this store of pleasure?"

Then the old man made answer. "Can I have any secrets from thee? Hearken, O king, and I will tell thee what lies behind the shows of the world. What I have shown thee is an outward sign and symbol of what thou shouldst do, but it lies far outside those recesses whither I shall lead thee. Thou couldst indeed give these beings a store of pleasure, and they would go through their actions until it was all spent; but then thou wouldst be as one of themselves. Thou wouldst have to perform the painful part of some action and let them perform the pleasant part, and thus thou wouldst be immersed in the same chain of actions wherein they were. For regard my staff as it begins to swing. It is not I that make the movement that is imparted to it; that movement lay stored up in my arm, and when I struck the staff with my arm it was as if I had let another staff fall which in its falling gave up its movement to the one I held in my hand."

"Where, then, does the movement go to when the staff ceases to swing," asked the king.

"It goes to the finer particles of the air, and passes on and on. There is an endless chain. It is as if there were numberless staffs, larger and smaller, and when one falls it either raises itself or passes on its rising to another or to others. There is an endless chain of movement to and fro, and as one ceases another comes. But, O king, I wish to take thee behind this long chain and to place thee where thou mayest not say, I will do this or that; but where thou canst say, This whole chain of movement shall be or shall not. For as thou regardest this staff swinging thou seest that it moves as much up as it does down, as much to right as to left. And if the movements which it goes through came together it would be at rest. Its motion is but stillness separated into equal and opposite motions. And in what thou callest rest there are vast movements. It shall be thine, O king, to strike nothingness asunder and make things be. Nay, O king, I have not given thee these beings in the valley for thee to move by outward deeds, but I have given them to thee such that thou canst strike their apathy asunder and let them live. And know, O king, that even as those beings are whom thou hast found, so are all things in the valley down to the smallest. The smallest particle there is in the valley lies, unless it were for me, without motion. Each particle has the

power of feeling pain and of feeling pleasure, but by the
law of the valley these are equal. Hence of itself no
particle moves. But I make it move, and all things in
the valley sooner or later move back to whence they
came. The streams which gather far off in the valley I
lead along to where they fall into the depths between
us. There they shiver themselves into the smallest fragments, and each fragment I cause to return whence it first
came. And, O king, in all this movement, since it ends
where it began, there is no more pleasure than pain. It
is but the apathy of rest broken asunder. But the particles will not go through this round of themselves. I
bear the pain to make them go through, each one the
round I appoint it."

"How then," exclaimed the king, thinking of the pain
he had felt in directing the movements of the children,
"canst thou bear all this pain?"

"It is not much," answered the old man; "and were
it more I would willingly bear it for thee. For think of
a particle which has made the whole round of which I
spoke to you—it will make this journey if on the whole
there is the slightest gain of pleasure over pain; and
thus, although for each particle in its movement at every
moment I bear the difference of pain, the pain for each
particle is so minute that the whole course of natural
movements in the valley weighs upon me but little.
And behold all lies ready for thee, O king. I have
done all that I can do. I can perfect each natural process, each quality of the ground, each plant and herb I
make, up to the beings whom thou hast found. They
are my last work, and into your hands I give them."

And when he had said this, the old man let drop his
staff, and placing both hands to his breast he seemed
to draw something therefrom, and with both hands to
fling it to the king.

For some moments' space the king could distinguish nothing, but soon he became aware of a luminousness over the mid ravine. Something palely bright was floating towards him. As the brightness came nearer he saw that it was a centre wherein innumerable bright rays met, and from which innumerable bright rays went forth in every direction.

"Take that," the old man cried. "The rays go forth unto everything in the valley. They pass through everything unto everything. Through them thou canst touch whatsoever thou wilt."

The king took the rays and placed them on his breast; thence they went forth, and through them he touched and knew every part of the valley. And thinking of the hut where the children lay, the king perceived through the rays that went thither that the walls were tottering, and like to fall on the children. And through his rays he knew that the children perceived this in a dull kind of way; but since in their life there was no more pleasure than pain, they did not feel it more pleasant to rise up and move than to be still and be buried.

But the king through the rays, as before through the reeds, took the pain of moving, and the children rose and came out of the hut; and soon they were with the king, running and bounding as never children leapt and ran, with ecstasy of movement and unlimited exuberance of spirit. But as they leapt and ran the king felt an increasing pain in all his limbs. Still he liked to see them in their full and joyous activity, and he wished them to cast off that dull apathy in which they lay. So all through the night he roamed about with them thinking of all the wildest things for them to do, and leading them through dance and play, every movement and activity he could think of.

At length the rising sun began to warm the air, and the king, exhausted with pain, left off bearing it for them.

After a few languid movements the children sank down on a comfortable bank into a state of absolute torpor. The king looked at them ; it seemed inconceivable that they could be the same children who had been running about so merrily a few moments ago. Thus far he had received no advantage from the rays the old man had given him, except that he could touch the children more easily.

He turned wearily and looked around. His horse stood there. But instead of whinnying and running up to greet him, the faithful animal stood still, looking across the ravine.

" Perchance without my burden, and with the strength these rays may impart," thought the king, " he might manage the leap."

The horse was standing opposite the remains of the natural bridge over which the two had so rashly crossed the day before. The king touched the horse with his rays. As with a sudden thrust of the spur, the noble animal rushed forward and leapt madly from the fragments of the arch. His fore feet gained the opposite brink, and with a terrible struggle he raised himself on the firm ground. Then he stood still. With a crash the remaining fragments of the bridge fell into the gulf, leaving the vast gap unnarrowed at any part. The horse stood looking over the ravine. But though the king called him by name, the faithful creature who used to come to him at the slightest whisper paid no heed. In a few moments he galloped off along the track the courtiers had pursued.

CHAPTER III.

THE king being left thus with the children, applied himself to thought. He directed his rays to one of the children and caused it to stand up, and, following the counsel of the old man, he thought of an action. The action he thought of was that of walking, and he separated it into two acts; the one act moving the right foot, the other act moving the left foot. And he separated the apathy in which the child was into pleasure and pain ; pleasure connected with the act of moving the right foot, pain connected with the act of moving the left foot. Immediately the child moved forward its right foot, but the left foot remained motionless. The child had taken the pleasure, but the pain was left ; or, since the king had connected the pleasure and pain with two acts, it may be said, had done the pleasant act and left the painful act undone.

After waiting some time to see if the child would move, the king took the pain of moving the left foot ; instantly the child moved it, and as soon as it had come to the ground again it moved the right foot, which was the pleasant act. But then it stopped. And by no amount of taking pains in the matter of the left foot could the king get the child into the routine of walking. As soon as he ceased to take the pain of moving the left foot, the child remained with the right foot forward. At last he removed his attention from the movement of the child, and it sunk back again torpor.

The rest of the day the king spent in reflection, and in making experiments with the children. But he did not succeed any better. Whatever action he thought of they went through the pleasant act, but made no sign of going through the painful act.

When darkness came the king perceived the faint luminousness of his rays : unless he had known of them he would hardly have perceived it.

And now he tried a new experiment. He took one of the rays, and, detaching it from the rest, he put it upon the body of one of the children, going out from its body and returning again to its body, so that it went forth from the child and returned to the child again. He then caused the child to stand up, and again tried it with the action of walking. His idea was this : the child required a power of bearing its own pain in order to go through a painful act, and as the rays enabled him to bear their pain, the ray proceeding from the child and coming back to it might enable it to bear its own pain. And now he separated the apathy into pleasure and pain as before. The child moved the right foot, and then when it had moved it, he saw that it actually began to move the left foot. But it did not move it a complete step, and after the next movement of the right foot the left foot did not stir.

Again and again the king tried the children, but his attempts came to nothing. One halting step of the left foot he could get them to go through, but no more.

He spent many hours. Suddenly the cause of his failure flashed upon him. " Of course," he said to him-self, " they don't move, for I have forgotten to take part of the pain. If they went on moving their left feet they would have no balance of pleasure."

And he tried one of them again. The child moved the right foot, then began to move the left foot. The king now by means of his rays took part of the pain of the movement of the left foot, and the child completed the step with it. Then of course it moved the right foot, for that was pleasant, and again the king took part

of the pain of moving the left foot, and the child completed its second step.   It walked.

The difficulty was surmounted.   Soon both the children were moving hither and thither like shifting
shadows in the night, and the king felt just a shade of
pain.

The children would come up to him and talk with
him, if he took the difference of pain which made it
pleasant for them to do so.   But they had no idea of
his action on them, for by his taking the difference of
pain they found an action pleasant, and felt a motive
in themselves to do it, which they did not in the least
connect with the being outside themselves to whom they
spoke.   They looked on him as some one more powerful than themselves, and friendly to them.

As soon as he was assured of the practical success of
his plans, the king let the children relapse into their
apathy while he thought.   He conceived the design of
forming with these children a state such as he had
known on earth—a state with all the business and affairs
of a kingdom, such as he had directed before.   The
vision of the palace which the old man had shown him
rose up.   He saw in imagination the fertile fields, with
the roads stretching between them ; he saw all the
varied life of a great state.   Accordingly from this time
he was continually directing their existence, developing
their powers, and learning how to guide them.   And
just as on first learning to read whole words are learnt
which are afterwards split up into letters by the combinations of which other words are formed ; so at first
he thought of actions of a complicated nature, such as
walking, and associated the moments of pleasure and
pain with the acts of which such actions were composed.
But afterwards he came to regard the simpler actions by
the combination of many of which the beings were made

to walk, and with the separate acts of these simple actions he associated pleasure and pain.

And at first the beings were conscious of these simple acts and nothing else, but in order that they might carry out more complicated actions, he developed the dim apprehension which they had, and led it on to the consciousness of more complicated actions. The simplest actions became instinctive to these beings, and they went through them without knowing why. But if at any time the king ceased to take the difference of pain, these actions, seemingly automatic as they were, ceased.

At certain intervals the king found his plans inconvenienced. Every now and then the beings went off into a state of apathy. Enough pain was borne for them to make it just worth their while to go through the actions of each routine. But any additional complication or hindrance unforeseen by the king was too much for them, and they sank under it. To remedy this he took in every action a slight portion of pain more than he had done at first. Thus he expended a certain portion of pain-bearing power to give stability to the routines. And the margin of pleasure over pain thus added was felt by the beings as a sort of diffused pleasure in existence, which made them cling to life.

Now in guiding these beings towards the end he wished to obtain, the king had to deal with living moving beings, and beings whose state was continually changing. And this led him to adopt as the type of the activity of these beings not a single action, but a succession of actions of the same kind, coming the one after the other. Thus a being having been given a certain activity, it continued going on in a uniform manner until the king wished to alter it.

Again it was important to keep the beings together, to prevent their being lost in the remote parts of the

4

valley, and consequently the king took, other things being equal, a certain amount of the pain of motion towards the centre, and took none of the pain in any movement away from the centre of the valley. Thus the inhabitants had a tendency to come towards the centre, for there was a balance of pleasure in doing so, and thus they were continually presenting themselves to his notice, and not getting lost.

Of course, if there was any reason why he wanted them away from the centre, the king ceased his bearing of the pain of motion towards the centre, and then they were under the other tendency solely, which he imparted to them, in virtue of his bearing pain in another respect. And in everything that he did the king had regard to the circumstances in which the beings were placed, and the objects which he wanted to obtain. He did not spare any of his pain-bearing power to give them pleasure purely as a feeling, but always united the pleasure he obtained for them by his suffering with some external work.

And as time went on and the number of the inhabitants increased, he introduced greater order and regularity into the numberless activities which he conceived for them. The activities formed regular routines, conditioned by the surroundings of the being and the routines of those around it. A routine did not suddenly cease without compensation; but if the king wished it to stop he let another activity spring up at once in place of it, so that there was no derangement. The beings gradually became more intelligent, so that they could be entrusted with more difficult routines, and carried them out successfully, the king, of course, always taking the difference of pain necessary to make it worth their while. And they even became able to carry out single activities on a large scale, involving the co-operation of many single routines. For they had a sense of

analogy, and observing some activity which the king
had led them through on a small scale, and in which
they had found a balance of pleasure, they were ready
to try a similar one on a larger scale.

There was one feature springing from the advanced
intelligence of the inhabitants which it is worth while to
mention. Many of the possible activities which the beings
could go through, instead of consisting of a pleasurable
part first and a less painful part afterwards, consisted of
a painful part first and a pleasurable part afterwards.
This might happen by the particular arrangement of the
acts of which the compounded activity consisted, the
acts having already moments of pain or pleasure affixed
to them, and happening to occur in such dispositions
that the first part of the activity was painful, the next
part pleasurable.

Now when the intelligence of the inhabitants was
developed, the king, by leading them to think of such
an activity, could induce them to go through with it.
For the idea of the pleasure which would accompany
the second part of the activity lightened the pain of
the first. And this, combined with the portion of the
pain which the king bore, almost counterbalanced the
pain connected with the first part of the activity. Thus
the beings were enabled to go through the painful part
of the activity. But when they came to the second part
of the activity the creatures were much disappointed. For
by the law of the valley pleasure and pain were equal
(except for the small part which the king bore). Now the
pleasures of expectation had been so great that when the
time came for the act usually associated in their minds with
pleasure, the pleasure due had most of it been used up.

From this circumstance a saying arose amongst the
inhabitants which was somewhat exaggerated, but which
had a kernel of truth in what has just been described.

The saying was that " The pleasure for which a labour
has been undertaken flies away as soon as the labour
has been finished, and nothing is left but to begin a new
labour." And, again, another saying : " The enjoyment
of a thing lies in its anticipation, not in its possession."

All this which has been so briefly described had in
reality taken a long time. And now fields were cultivated, better houses were built. The inhabitants of the
valley had increased greatly in number, and were divided
up into several tribes, inhabiting different parts of the
valley. But the most favoured position was the centre,
and for the possession of the centre there were con-
tentions and struggles. There the king's activity in
bearing was greatest, and the life was most developed.

All around the outskirts of the valley dwelt the ruder
and less advanced people, who were called barbarians
and savages by those nearer the centre.

CHAPTER IV.

Now when the king saw the inhabitants becoming more
like the human beings he had known, he felt that he
was solitary, and he desired to have some intercourse
with them. But when he appeared amongst them they
recognized him at once as some one more powerful than
themselves, and were afraid of him. In their alarm they
tried to lay hands on him. When he, to prevent their
attacks, withdrew his continued bearing the difference of
pain in their actions, those who were attacking him
sank into apathy and became as the children whom he
had first found.

And a horrible report sprang up amongst the inhabitants of a terrible being who came amongst them, and

who struck all who looked on him with torpor and death. So the king ceased to walk amongst them. Still it was long since he had heard the sound of a voice speaking to him, and he wished for a companion. He sought again the old man, and standing at the edge of the chasm he called upon him.

And the old man appeared. "Art thou weary, O king, of thy task?"

"Nay," replied the king; "but I wish to make myself known to the inhabitants that I may speak with them and they with me."

And the old man counselled him to give some of his rays to one amongst the beings, for then this being having these rays and the power of bearing pain for another other than himself, would be like the king, and being like him would understand him.

Now the king sought over the whole of the valley, and of all the inhabitants he found one most perfect in form and in mind. He was the son of a king, and destined to reign in his turn over a numerous people. And the king gave him some of his rays, straight rays going forth from the prince to others.

And immediately the prince awoke as it were from a dream. And he comprehended existence, and saw that in reality the pain and the pleasure were equal. And when he had seen this, and knew the power of the rays, and how by bearing pain he could make others pass through pleasure and pain, and call those sleeping into activity; when the prince knew this, he cried out:

"One thing succeeds another in the valley; pain follows pleasure, and pleasure follows pain. But the cause of all being is in bearing pain. Wherefore," he cried, "let us seek an end to this show. Let us pray to be delivered, that at last, pain ceasing, we may pass into nothingness."

Thus the prince, apprehending the cause of existence, felt that it was pain, and dimly comprehending how the king was bearing pain, and himself feeling the strenuousness of the effort of using the rays for which the frame of the inhabitants was unstrung, longed that existence itself might cease.

Yet all his life his deeds were noble, and he passed from tribe to tribe, bearing the burdens and calling forth the sleeping to activity.

CHAPTER V.

It is now the place in which to give a clear account of the king's activity, and explain how he maintained the varied life of the valley.

And the best plan is to take a typical instance, and to adopt the Arabic method of description. By the Arabic method of description is meant the same method which the Arabs used for the description of numerical quantities. For instance, in the Arabic notation, if we are asked the number of days in the year, we answer first 300, which is a false answer, but gives the nearest approximation in hundreds ; then we say sixty, which is a correction ; last of all we say five, which makes the answer a correct one, namely, 365. In this simple case the description is given so quickly that we are hardly conscious of the nature of the system employed. But the same method when applied to more difficult subjects presents the following characteristics. Firstly, a certain statement is made about the subject to be described, and is impressed upon the reader as if it were true. Then, when that has been grasped, another statement is made, generally somewhat contradictory, and the first notion formed

nas to be corrected.   But these two statements taken
together are given as truth.   Then when this idea has
been formed in the mind of the reader, another statement is made which must likewise be received as a
correction, and so on, until by successive statements
and contradictions, or corrections, the idea produced
corresponds to the facts, as the describer knows them.

Thus the activity of the king will be here described
by a series of statements, and the truth will be obtained
by the whole of the statements and the corrections
which they successively bring in.

When the king wished to start a being on the train
of activity he divided its apathy into pleasure and pain.
The pleasure be connected with one act which we will
call A.   The pain he associated with another act which
we will call B.

These two "acts," A and B, which together form what
we call an "action," were of such a nature that the
doing of A first and then of B was a process used in the
organization of the life in the valley.

Thus the act A may be represented by moving the
right foot, B by moving the left foot, then AB will be
the action of taking a step.   This however is but a
superficial illustration, for the acts which we represent
by A and B were fundamental acts, of which great
numbers were combined together in any single outward
act which could be observed or described.

Suppose for the present that there is only one creature
in the valley.   The king separates his apathy with
regard to the action AB.   Let us say he separates his
apathy into 1000 pleasure and 1000 pain.   Of the
pleasure he lets the being experience the whole, of the
pain he bears an amount which we will represent by 2.
Thus the being has 1000 pleasure and 998 of pain, and
the action is completed.   His sensation is measured by

the number 1000 in the first act, and by 998 in the second act.

But the king did not choose to make the fundamental actions of this limited and finishing kind. As the type of the fundamental activity, he chose an action, and made the being go through it again and again. Thus the being would go through the act A, then the act B. When the action AB was complete it would go through an act of the kind A again, then through an act of the kind B. Thus the creature would be engaged in a routine of this kind, AB, AB, AB, and so on.

And if the creature had been alone, and this had been the sole activity in which it was concerned, the king would have gone on bearing 2 of pain in each of these actions. The king would have kept the routine going on steadily, the creature bearing 1000 of pleasure in each A, and 998 of pain in each B.

At this point it may be asked that an example should be given of one of these elementary routines which the king set going. And this seems a reasonable request, and yet it is somewhat too peremptory. For in the world we may know of what nature the movements of the atoms are without being able to say exactly what the motion of any one is. In such a case a type is the only possible presentation. Again, take the example of a crystal. We know that a crystal has a definite law of shape, and however much we divide it we find that its parts present the same conformation. We cannot isolate the ultimate crystalline elements, but we infer that they must be such as to produce the crystal by their combination.

Now life on the valley was such in its main features as would be produced by a combination of routines of the kind explained. There were changes and abrupt transitions, but the general and prevailing plan of life was that of a routine of alternating acts of a pleasurable and a painful kind. It was just such as would be built up out of elementary routines, on which the king could count, and which, unless he modified their combinations, tended to produce rhythmic processes of a larger kind. And even the changes and abruptnesses had a recurrent nature about them, for if any routine in the valley altered suddenly, it was found that there were cases of other routines altering in like manner, when the conditions under which they came were similar. Thus the fundamental type of the action which the king instituted was that of a routine AB, AB, as described above. But there were two circumstances which caused a variation, so that this simple routine was modified.

Firstly, there was not one being only but many.

Secondly, the king wished to have some of his pain-bearing power set free from time to time. He did not wish to have to be continually spending it all in maintaining the routines he had started at first, and those immediately connected with them.

When he first began to organize the life of the beings he did not consciously keep back any of his pain-bearing power, but threw it all in the activities which he started. Still from time to time he wished to start new activities quite unconnected with the old, and for this reason he withdrew some of his pain-bearing power, as will be shown afterwards.

There were many beings. The king chose that the type of activity in each should be a routine. In that way he could calculate on the activity, and hold it in his mind as a settled process on whose operation he could count. But as the routines of the beings proceeded they came into contact with one another, and made, even by their simple co-existence, something different from what a routine by itself was. They interwove in various

ways.   Then, in order to take advantage of the combinations of these routines, or to modify them, it was
necessary to set going other routines.

In order to be able to originate these connected
routines the king adopted the following plan.

In the first action AB he separated the creatures'
apathy into 1000 pleasure and 1000 pain, bearing 2 of the
pain himself.   The creature thus went through 1000 of
pleasure and 998 of pain.   In the next action AB he did
not separate the beings' apathy up into so much pleasure
and pain.   He separated it up into 980 pleasure and 980
pain, that is, each moment of feeling was 20 less in sensation than the moments of feeling were in the first action.

Now it is obvious that if the bearing 2 of pain will
make it worth while for a being to go through 1000
pleasure and 998 pain, then the bearing on the king's
part of 1 of pain would make it worth while for the
being to go through 500 pleasure and 499 of pain.

And a similar relation would hold for different amounts
of pleasure and pain.   Thus clearly for the being to go
through 980 of pleasure and the corresponding amount
of pain, it would not be necessary for the king to bear
so much as when the being went through 1000 of
pleasure and the corresponding amount of pain.

Consequently when the king divided the beings'
apathy into 980 pleasure and 980 pain, it would not be
necessary for him to bear 2 of pain to make it worth
the beings' while to go through the action.   The king
would not bear so much as 2 of pain, and thus he
would have some of his pain-bearing power set free.
He would have exactly as much as would enable him to
make it worth a being's while to go through an action
with the moments of 20 of pleasure and 20 of pain.

And this—with a correction which will come later—
is what the king did.   He employed the pain-bearing

power thus set free in starting other routines. Thus in the routine AB, AB, AB there would be first of all the action AB. Then along with the second action AB, the king (with the pain-bearing power set free) started an action CD—the beginning of a routine CD, CD, CD. Thus as the first routine went on and came into connection with other routines, new and supplementary routines sprang up which regulated and took advantage of the combinations of the old routines.

The amount of the moments of pleasure in the routine CD, was (with a slight correction explained below) measured in sensation, equal to 20. Thus the moment of pleasure in the first A being 1000, the moment of pleasure in the second A was 980, the moment of pleasure in the first C was 20 (subject to the correction spoken of). Thus the total amount of sensation in the second A and and the associated act C, taken together (but for a small correction) was equal to the sensation in the first A. Hence the three points which were characteristic of the activity of the beings in the valley are obvious enough.

1. There is as fundamental type a routine AB, AB. AB, the sensation involved in which goes on diminishing.

2. There are routines CD, CD, &c., connected with AB, AB, in which the sensation which disappears in the routine AB, AB seems to reappear.

3. In the action AB itself there is a disappearance of sensation. The sensation connected with A is 1000, that connected with B is 998. Thus 2 of sensation seems to have disappeared. This 2 of sensation is of course the pain which the king bore, and which was the means whereby the creature was induced to go through the action at all. But looked at from the point of view of sensation, it seems like a diminution of amount. This diminution of amount, owing to the correction spoken of above, was to be found regularly all through the routine.

And now, with the exception of the final correction, the theory of the king's activity is complete. There are certain mathematical difficulties which render an exhaustive account somewhat obscure in expression. When we take a general survey of a theory we want to see roughly how it all hangs together ; but if we mean to adopt it, the exactitude of the numerical relations becomes a matter of vital importance.

It must be added that the numbers taken above were taken simply for purposes of illustration. In reality the pain born by the king was less in proportion.

The exhaustive account which follows deals with small numerical quantities. It had better be omitted for the present, and turned to later on for reference.

EXHAUSTIVE ACCOUNT.

We keep for the time being to the numbers used above. When the king had enough pain-bearing power set free in the second action of the routine AB, AB to start another routine CD, of 20 pleasure 20 pain, he did not use it all. He only used enough of it to set a routine going the moments of pleasure and pain in which were 16 in sensation. The routine CD was made up of acts with 16 of pleasure and 16 of pain.

The sensation in the first A was 1000, in the first B it was 998, giving a disappearance of 2. In the second A it was 980, and in C, which starts concurrently with the second A, it was not 20 as might have been expected, but 16, giving a loss of 4. The second A is less than the first A by 20. Searching for that 20 we find 16 in C. But there has been a disappearance of 4.

Looking now at the successive acts in the series we have in A 1000 sensation, in B 998 sensation, in A and C together 996 sensation.

The cause of the loss between A and B has already been explained. That between B and the second A with C remains to be accounted for.

It has been already said that the king withdrew some of his pain-bearing power from the routine AB and all routines connected with it, thus he was enabled to start activities altogether unconnected with those which he had originated, and was with regard to the products of his own activity as he had been at first, with regard to the beings in the valley before he started them on the path of life. And it was in consequence of his withdrawal of his pain-bearing power that the amount of sensation in C was not 20 but was less. This loss of 4 of sensation to the being corresponded to a setting free of a certain portion of pain-bearing power on the part of the king. And thus as the process went on, a portion of his power was continually being returned to him.

In the table below the first line of figures contains the amount of sensation in the actions AB, AB. The second line of figures contains the amount of sensation in the actions CD, CD. The third line of figures relates to another connected routine EF, EF, which originates in a manner similar to CD. The fourth line of figures represents the amount of pain borne by the king, the fifth line represents his pain-bearing power set free.

(1)	1000	998	980	$978\frac{40}{1000}$	960	$958\frac{80}{1000}$
	A	B	A	B	A	B
(2)			16	$15\frac{998}{1000}$	$15\frac{630}{1000}$	$15\frac{658640}{1000000}$
			C	D	C	D
(3)					16	$15\frac{998}{1000}$
					E	F
(4)	2			$1\frac{992}{1000}$		$1\frac{991}{1000} + \frac{360}{1000000}$
(5)	0			$\frac{8}{1000}$		$\frac{8}{1000} + \frac{640}{1000000}$

If the total amount of sensation which is experienced by the being in the original routine and the connected routines in the consecutive stages be summed up, it will be found to be

$$1000, \; 998, \; 996, \; 994\tfrac{38}{1000}, \; 991\tfrac{680}{1000},$$

and so on.

Finally, the proportion of pain borne by the king was so small compared with the sensation experienced by the being, that A and B were apparently equal in sensation. Thus the sensation in the second A and in C together becomes apparently equal to that in B. And instead of the sensation diminishing quickly as shown above, it was only after a great many acts of the primary and connected routines had been gone through that any diminution of sensation in the form which the being could experience it was to be detected   Thus, as before stated, there was:

1. A routine of continually diminishing sensation.

2. Connected routines the sensation in which was apparently equal to that lost in A.

3. There was a continuous disappearance of sensation from the experience of the beings accompanying every step of the routine. The sensation which they could experience was less in every subsequent step and connected steps than in any one in which it was measured.

CHAPTER VI.

THE history of the events which took place in the valley in their due order and importance must be sought elsewhere. But let us return and look at the condition of the valley and its inhabitants. Let us see what has become of them after a great lapse of time.

It is a fair, a beautiful land. The greater part of it is cultivated. There is no war—even to the extremest confines of the valley there is peace. Passing from the remote confines where still dwell a barbarous race, we come, as we approach the metropolis, amongst a more and more polite and refined people. In the metropolis itself the buildings are numerous and of great size. The palace which the king saw rise under the old man's music is there, but another ruler dwells in it. Near the palace are two vast buildings standing on each side of a wide open court. There is no other building near save one between them, a comparatively small edifice of brick. These buildings are the assembly halls of the two most important councils in the valley. In the one on the left-hand side of the palace met the most distinguished of the inhabitants who from a special inclination or fitness were entrusted with the regulations about the pleasure and pain of the inhabitants. They framed the rules according to which each inhabitant must conform in his pursuit of pleasure, and they made the regulations whereby the whole body of inhabitants were supposed to gain an increase in pleasure and to avoid pain.

In the building on the right hand of the palace met those of the inhabitants who had studied the nature of feeling most deeply, and who from temperament or for other reasons had in their course of study not paid so much attention to whether feelings were painful or pleasurable, but who had studied their amount and regularity of their recurrence. They were the thinkers from whom all the practical inhabitants derived their rules of business. They devised the means and manner of putting into execution what was decided on in the other assembly. They did not often propose any positive enactment themselves, but were always able to show

how the proposals of the other council could be carried into effect.

Their power was derived in this manner. The king had connected the feelings of pleasure and pain with certain acts, and had given each being a routine. Now as he himself made use of this routine and combined the routines of different individuals to bring about the results he desired, so also did the rulers of the valley. The routines of the individuals were studied and classified, and if any work was required to be done, those individuals whose routines were appropriate were selected and brought to the required spot. Now to effect this a careful study of the different routines was necessary, and also a knowledge of what stage they were at. For it would be no use bringing an individual whose routine was almost at an end to a work which was just beginning. Hence the most delicate instruments and processes had been devised for measuring the amount of feeling experienced by any individual, whether of pleasure or of pain, and a careful classification had been made of all routines.

But it is best to study the constitution of the state in a regular order, and the questions of pleasure and pain considered as such were esteemed the most important.

The inhabitants knew that they sought pleasure and avoided pain, and the great object was to make their life more pleasurable. Two means were adopted, the banishing of the causes of pain, and the obtaining causes of pleasure.

By causes of pain and pleasure they meant those objects with which the king had associated the feelings of pleasure and pain in the equal and opposite moments into which he had divided their apathy.

But in this respect they were in error to a certain extent, for it was not so much in respect to things as in respect to actions that the king separated their apathy into pleasure and pain. For instance, there was a peculiar species of shell which was found in many parts of the valley, covered with strange and involved lines and marks. Now the king had struck the apathy of the inhabitants into two moments with regard to this shell, one of pain connected with tracing out the twistings and interweavings of the hues on the shell, one of pleasure in contemplating the shell when the twistings and inter-weavings had been deciphered. Now it was the custom of the inhabitants to call the shell in its undeciphered condition a painful object, in its deciphered condition a pleasant object. And whoever could, would get as many deciphered shells as possible and experience the wave of pleasure in looking at them.

Now in the earlier ages those who deciphered the shells, or did work of a similar kind, had been forced to do it ; they were a kind of slaves dependent on the will of their masters, who took away all the pleasures of their life. But in these earlier ages a great danger arose, for when all the pleasure was taken away by their masters, great masses of these slaves sank into apathy, and it seemed as if the valley was sinking into deadness.

Now this was a great terror with the inhabitants whose life was pleasurable, and at length they determined that there should not be any more of these slaves. But each of the inhabitants when he worked for another had to have it made worth his while.

In this way a great diminution took place in the pleasure-giving power of the so-called pleasurable things. For if a man had had it made worth his while to decipher one of these shells, he had had a great deal or nearly all of the pain he spent in doing it counterbalanced by the pleasure given him to induce him to do it. Hence when the shell was handed over there was not much to enjoy

5

in it ; for by the law of the valley the pleasure and the pain were equal, and the decipherer, not having gone through so much pain on the whole, there was but little pleasure to be got.

In fact, at this time the fashion of filling the houses of the more powerful of the inhabitants with the so-called pleasurable things had somewhat gone out, and it had passed into a proverb, "It is better to decipher your own shells."

Now it may be considered strange how it was that some of the inhabitants could get other of the inhabitants to decipher the shells for them at all, or, at any rate, to decipher them so that there was any balance of pleasure left with the shells at all. But this power on the part of some of the inhabitants depended on the general action of the king. For by bearing the difference of pain in innumerable respects in the life of each he made life a pleasure (on the whole) to each, and they strove each to preserve their own life which was a source of pleasure. And some of the more powerful inhabitants had the power of denying to the rest, unless they laboured for them, the means of continuing to exist. Consequently it was possible for things to be obtained by the more powerful which had a balance of pleasure in them.

But the authorities who had studied the life of the valley in relation to pleasure and pain, saw that there was a danger in this relation of the more powerful to the less powerful. For as the numbers of the inhabitants increased the power grew more and more concentrated in the hands of a few, and there was a tendency for the inhabitants in general to be compelled more and more to go through the painful part of actions, leaving the pleasurable parts for the more powerful. And every now and then, before the council of wise men regulated the matter, great masses of the inhabitants passed off in a state of apathy. So they had many laws to restrict the action of the more powerful of the inhabitants; and, indeed, the more powerful of the inhabitants were ready to frame these laws themselves, and were willing to obey them, for they did not like to see portions of the inhabitants going off into a state of apathy.

But not only in this respect, but also in every other, the wise men regulated the affairs of the valley so as to make life more pleasurable. They had severe laws against any one who deprived another of pleasure without his consent, by violence or deceit. They did all they could to ward off a state of apathy. But in one respect beyond all others they were full of care and precaution. And this was in guarding against such sources of trouble, anxiety, and pain which could be removed from the community as a whole. Anything tending to lower the standard of comfort as a whole was carefully removed. Irregularities were reduced as much as possible; and, in one respect, a great step had been taken. It had not been carried in the council of wise men without great opposition, but it had at length been passed into law.

Any child born in the valley which had any incurable disease, or any gross deformity, or which by its delicacy seemed likely to cause more pain than pleasure in the valley, was at once put out of existence. The gain to the inhabitants of the valley of this was in their eyes immense; for their sight was offended by no deformities, and the painful offices of attending to the sick had undergone a considerable diminution since this edict had been passed into law.

The important duty of deciding on the claims of every infant that was born to a painless extinction was confined to a band of inspectors, who stayed for a short time only in any one part of the valley, lest they should become biassed by personal acquaintance with the individuals for whose children their offices were called into requisition

---

CHAPTER VII.

PASSING on to the other great building, where the other wise men meet, it is right to describe what may be called the intellectual development—as the foregoing was the moral development of the valley. The course which the opinions of the thinkers in the valley had gone through was the following.

At first they had no clear ideas, but all manner of mere opinions and fancies. At last they apprehended certain general tendencies—such as that towards the centre of the valley, and they explained many inclinations which had before been puzzling to them by this. And stimulated by this great discovery they examined more and more closely. And they found many special tendencies like that towards the centre of the valley, which the king had called into existence, and which he let go on as a general rule, unless he wished the contrary. And they also succeeded in nearly isolating the simplest routines, and so practically could observe the type of the king's plan.

They saw that one act A was succeeded by another act B. And not taking into account that one was pleasant the other painful, they measured the amount of sensation present in the two acts. And then they took the next pair of acts, namely, A and B over again, and measured the amount of sensation present in them ; and they found that the amount of sensation gradually diminished. And at first they thought that sensation gradually came to a stop; but afterwards they noticed that other actions were started in the neighbourhood of the routine A B as that diminished in point of feeling.

Now, of course, these other actions were started by the king with the pain-bearing power set free from the routine A B, as above described. But not knowing anything about this action on the part of the king, or about the king at all, the conclusion arrived at was this : That sensation transmits itself. If it does not continue in the routine A B, that part which does not continue passes on to the other routines, C D, E F, &c.

Then they measured very carefully; and they found, as nearly as they could measure, the routines which sprang up as the routine A B died away were equal in sensation to the loss in the routine A B, A B. And from this they concluded that the amount of sensation or feeling was constant. They called it living force, and said that it must transmit itself and, wherever it appeared, be equal in its total amount to what it was at first. But after a time, with more delicate measurements and more patient thought, they found that some of the sensation was still unaccounted for.

For consider any routine consisting of the acts A, B; A, B; A, B. In order to make any pair of acts A, B worth while, the king bore a certain amount of pain. Referring to the numbers which we took before, if there were 1000 of pleasure in A there would only be 998 of pain in B. Thus the sensation was not equal in the two acts A and B. Some of the sensations had gone, and the portion of sensation we are now considering—the portion by which B was less than A—had not gone in starting other routines. This loss could not be accounted for as they accounted for the difference in sensation between the first action A B and the second action, consisting of the acts A and B in the routine.

There was a loss of sensation which was counterbalanced by the gain in sensation in other routines.

But besides this there was a further loss. Some
sensation went off, not to be recovered in any routine
they knew.

Now it was the bearing on the part of the king which
produced the appearance of the passing away of sensation altogether, so that the act B was less in amount of
sensation than the act A. But the inhabitants—at least
the wise ones—were firmly convinced that sensation
could not be annihilated or lessened. So they came to
the conclusion that sensation was passing off into a form
from which it never reappeared, so that it could affect
them. They conceived it still to exist, but to be irrecoverably gone from the life of the inhabitants of the
valley.

Taking the numbers we have taken, and the simple
instances we have supposed, this course of reasoning
appears straightforward enough. But in reality so complicated was the state of things in the valley, and the
proportion of pain which the king bore in each single
action so minute, that to have arrived at this result
implied powers of investigation of no mean order.

It is interesting to mention the names which these
investigators gave in the valley. In the performance of
the pleasant act A, they said that the being acquired
greater animation. In going through the painful act B,
they said that he passed into a position of advantage.
They used the term advantage because, having completed
the painful act B, he was ready to begin the pleasant
part of the action A over again. And in this part he
manifested more animation.

And now although acts of greater animation and
greater advantage succeeded one another, and although
the new total of the sensation in the act of a being was

very nearly equalled by that in a subsequent act, still there was not—they had to confess there was not—a complete equality. Some of the sensation had certainly gone from the sphere in which the inhabitants could feel it.

We see that this sensation which was gone was in reality the pain-bearing of the king, which set all their life going.

But they knew nothing of this, and formed a very different conclusion. They said : " If some of the sensation is continually going and disappearing from the life of the inhabitants of the valley—if this is the case, although the sensation may not be destroyed, it is certainly lost to us."

And then they thought : " Surely the amount of sensation must be always the same ; if some of it continually goes off into a form in which we cannot feel it, that portion which is left behind, and which we feel, must be continually growing less."

Hence they concluded that the sensation in the valley was gradually running down. Less and less was being felt. After a time, which they calculated with some show of precision, all feeling will have left the inhabitants and gone off in some irrecoverable form. All the beings of the valley will sink into apathy.

Thus coming in the course of their investigations upon the action of the king, which was the continual cause of all life, they apprehended it as the gradual annihilation of life.

The small building between the two council halls remains to be noticed.

Now when the king had connected pleasure and pain with different acts to be performed by the inhabitants of the valley, he had of necessity to let the pleasant one be the one that came first in the order of its possible per-

formance. And then by the device of the curved rays
he had brought it about that the inhabitants went
through the painful act consequent on the pleasurable
one, the two together forming the complete action which
the king had designed. But this chain was not very
secure. The inhabitants had a tendency to go through
the pleasurable act and leave the painful act undone.

Now in things which necessarily concerned their life,
the king would, by repeatedly bearing the pain of the
painful act, continually set the beings going again ; for
when they had performed the pleasant act they were
landed in a state of torpor, until the pain of the painful
act had been borne by them or for them. Now if this
act of which they took the pleasant and left the painful
part undone was in the main current of their lives, the
king would over and over again, by bearing the pain,
bring those who had shirked the painful part into a
position of advantage again, so that they could begin
the routine afresh with another pleasant act. And often
when thus started again they would take to the routine,
and bear the pain in the painful act themselves. But
many, after assisting them again and again, the king
was obliged to let sink into apathy, such namely as
always left the painful part of the action undone.

Now the little building was the council hall or investigation chamber of the searchers out of new pleasures.
And by new pleasures they meant something of the
following kind. With the pleasant and painful acts
which made up the main routines of their life, it was not
safe to take the pleasant and leave the painful acts, for
that gradually led to their sinking into apathy. But
there were many routines which the king had instituted
besides the main ones. And if the pleasant part of the
action constituting these secondary routines were taken,
then there followed no tendency to lethargy in the main

current of their lives, but they simply had a pleasure the more. Of course the pain of the painful act had to be borne, but they not going through with it left it for the king to bear.

Long ago, through one of the inhabitants of the valley with whom he had communicated, the king had sent a message, asking the inhabitants not to take the pleasant part of an action without the connected painful part. But now all memory of this message was lost, and the little building had been built, as a council hall or investigation chamber for the searching out of pleasurable acts. In it all possible novelties of action were discussed. And the pleasant parts of them being described, exactly how far they were pleasurable, and in what degree they were pleasurable, the information was made public throughout the land.

CHAPTER VIII.

BESIDES these two principal buildings in the metropolis, there were other public buildings devoted to various purposes. And some of the most important were colleges devoted to the education of the young inhabitants.

Now there was in the college of applied sensations a student who, though outwardly as proficient as the average of his companions, was in reality the most backward of all. He learned by a kind of rote all the doctrines they understood, and he could explain apparently how one feeling caused another. But in himself he had no particle of understanding. He seemed deficient in the sense of cause and effect which the others had. Of this the following instance will suffice to show the nature.

The king had, in order to prevent the inhabitants from straying too far from the metropolis, kept a constant watchfulness over their movements, and had uniformly taken somewhat of the pain from any effort which they made to move towards the metropolis, and had not taken any of the pain in efforts whose tendency was to remove them to a distance from the metropolis. If there was any purpose to be served in going away from the metropolis, he took enough pain from these movements to make it worth the beings' while to go away from the metropolis. But when other things were equal, it was a pleasurable thing to go towards the metropolis. The king made this general inclination, because if it had not been so, beings lying out of the way of his immediate attention might have drifted away and gone to the confines of the valley, away from where the busy life he was calling out was manifested, and so have been lost to others and themselves. As it was by imparting this general pleasurableness of moving towards the metropolis he held all the inhabitants together, and knew the direction in which each would tend, unless for any special reason he had made it more pleasurable for the person to move away from the metropolis.

Now, as has been mentioned above, this general tendency had been observed by the inhabitants ; and they knew quite well that every individual tended towards the metropolis, and was only prevented from coming into it by strong local interests, or by all available positions in it or near it being already occupied. If any situation was vacant in the metropolis, it was easily filled up by those from the surrounding country, for they all felt this tendency to press in.

Now, the learned men in the valley had long recognized this as one of the most important laws of the valley. And tne students in the college of applied sensations felt this law to be true law, and anything which followed from it they felt to be self-evident. But the student of whom we speak had not this happy, settled feeling with regard to this law. He could not feel as if it were necessarily true.

One day the head of the college was talking to the foremost students—those who had nearly finished their course and who would take their places in the valley shortly—and he said incidentally in the course of his remarks, that those who were moving away from the metropolis were as much attracted to it as those who were moving towards it.

"Why do they move away, then?" asked the backward student, who had by great diligence, after a long time, plodded his way by force of remembering by heart into the top class. He forgot his usual caution and his acquired habit of only asking questions he had heard asked before in order to refresh his memory with the answers he had heard given before.

The professor frowned at the stupid question. "The supposed being," he answered, "while he is attracted to the metropolis in accordance with the general law, may yet have some stronger inducement at the time to move away from the metropolis. That he does move away shows of course that his temporary inducement to move away is stronger than his permanent attraction towards the metropolis."

The student said that he was obliged for the explanation. "But—— "

"Well?" said the professor.

"The only reason you have for supposing that the being is attracted towards the metropolis is that he does move towards the metropolis. I don't see why you should say it was pleasant for him to move towards the metropolis when he does not do so."

" But we know," said the professor.

" No," said the student, " you only suppose ; because you find it so on a great many occasions, you suppose it is so always. You are like a savage who attacks the house of a civilized man. And he tries the window, the civilized man meets him there ; so he tries the door, the civilized man meets him there ; so he goes back to the window, and is met there again. And he concludes there are two men in the house ; and after a time he concludes there are as many men in the house as there are ways by which he tries to get in."

The student had forgotten himself in speaking like this ; and the comparison to a savage, though made in haste and in good part as an illustration, offended the professor, so he said :

" You do not believe that the law of attraction towards the metropolis is universal, and affects all the inhabitants ? "

" I cannot," said the student.

" Then you shall go to a place where you will feel it," said the professor. " You will go to-morrow to the extreme confines of the valley, and stop there until you are of a different mind."

He said this in a superior and gentle manner. But it was a terrible blow to the prospects of any student to be thus exiled. And yet the professor was within his strict legal right, and the student knew it. He had avoided this danger all through his college course, and now it came with crushing effect on him. For just as long ago in the valley they had had doctrines about the king, and had punished any one who did not feel them as true, and who was found out, so now when all the ideas about the king had been disproved, they had severe regulations about the belief in the laws. The learned class was a sect of priests, and whoever threatened to bring confusion and trouble by denying any of the known laws, and to lead the ignorant people to disregard them and deny them, was subject to severe punishments. In the case of this student, the error did not so much matter, because he had committed his offence in the presence of well-instructed people, who would only smile at his folly. But he had in his presumption insulted the head of the college, and his punishment was universally considered to be mild and just. And yet he was not altogether in the wrong. For it was not as though the king (when he wanted a being to move away from the metropolis) took as usual a portion of his effort in going there ; and at the same time counterbalanced this by taking a still larger portion of the pain involved in his moving away from the metropolis. By no means. When the king willed a man to move away from the metro-polis, he let him start afresh, as it were, according to the conditions which every being was subject to in the valley—that it was just as pleasant as painful to move in any way, and he took a portion of the pain involved in moving away from the city.

Now the student, when he was sent away, tried ear-nestly to see wherein he had been wrong. The place where he was exiled was on the confines of the valley, where a peaceable race of savages lived, engaged in agriculture. In the quiet, monotonous life of the place he thought over his whole course of life, but could not obtain any different feeling. And while thus buried in thought, he fell into the way of going about with the savages and doing as they did. Much to his surprise, when his preoccupation of mind passed away, he found himself singularly at home with them. Their tastes seemed to agree with his. And he came to the conclu-sion that he was in reality a savage who by some mis-take had been admitted to the college. Having formed this conclusion, he threw himself into the life around him heartily. In course of time he won the confidence of the rude, uncultivated people, and they talked to him unreservedly.

Many curious traditions were handed down amongst them. There were some which proceeded from the time when the king had walked and talked with the children he called into activity. There were others proceeding from times when there had appeared amongst them one to whom the king had given some of his rays, so that that person had the power of making the pain less in actions for others, and of giving them motives to act, and of rousing them thus to an active state. And all these traditions they told to the exiled student.

Now their own belief was this. They thought that there was a power over them, and in this they recognized the king; but how it was that this power prompted them they did not know. Yet they connected him in some way with pleasure and pain. They thought it pained him when they had pleasure, but not in the way in which was really the case. They thought simply that it was pain to him to see them taking pleasure. They thought, moreover, that he would, if they displeased him much, take away all their pleasure and leave them nothing but pain.

Now the student saw clearly some errors, some contradictions in their belief. For instance, he knew that beings only followed pleasure, and directly pleasure was equalled by pain, sank into apathy, and then gradually vanished away. Hence he knew there need be no apprehension of the power's acting as they thought. But the thing they said, that their taking pleasure pained this power, struck him. He did not approve the results in their life, for it was in consequence very gloomily framed, though with a good deal of unconscious cheeriness. But he knew as a scientific fact that there was a constant diminution of feeling ; and since he also knew that beings in the valley did nothing except it was more pleasant, he concluded that although pleasure and pain might both be disappearing, still pain must be disappearing to a greater extent. Now since the feeling did not become nothing, but passed away out of the perception of the inhabitants, it followed that it must pass to some being. It did not disappear as feeling, but passed away from the sensation of the inhabitants. Is there a being, then, he asked himself—the power of whom these simple folks tell—who bears the difference of pain, and so makes existence pleasant to us ? And is that the meaning of what they say that our pleasure pains him ? Is it just the truth read backwards—the truth, namely, that by his taking pain we have pleasures, which they have had handed down to them as this—that our taking pleasure pains him.

When he had thought thus far he remembered one of his books in which the ancient beliefs of the valley were discussed. It happened to be one of the books which he had brought into his exile with him. He took it down, and in the evening set himself to search through it. And in a footnote towards the end of the book he read :

" The existence of a power shaping the valley for the good of the beings in it is clearly disproved. First, by the amount of suffering there is in the valley. Secondly, by the fewness of the types of life, and the constant modification of one plan to secure different results— which would be much better achieved by the use of radically different types and means. Thirdly, by the absence of any indication of such a power, except in the traditions of uncultivated tribes."

When the student had read this he rose up and paced his chamber. For he saw clearly that if it was in bear-

ing part of tne pain that the power of the being lay, the
first of these arguments fell to the ground. The presence
of the pain in the valley would prove that this power
took only some of the pain and not all. As to the
second argument, all it would come to was that the
being who, bearing pain, gave existence to the inhabitants, used economy in his actions—he chose to effect
his objects with the least possible expenditure of means.

Reflecting thus he went out.

Now it may be considered surprising that the king
did not communicate in some way with the student, for
by means of his rays he was in possession of all that
had gone on in his mind. But the king had found over
and over again that if he manifested himself to any one
of the inhabitants of the valley, the effect, though good
at the immediate time, was most disastrous for the following time. For the ends he was working towards,
and leading the inhabitants towards, were much greater
than any one of them could grasp or conceive. And the
inhabitants, as soon as they had communication with
him, at once thought they knew his final will. And
they were a set most peculiarly stiff in their notions, and
with the kind of sanction which communication with
him gave them, even the most absurd ideas if once conceived took a very long time to eradicate.

So when the student went out into the open air he
saw nothing except the stars, and heard nothing except
the wind. The way was so well known to him, however,
that he walked on quickly without stumbling in the darkness. He had not gone far when he saw a kind of
luminousness. Is it the moon beginning to rise? he
thought. But he found he had passed the light and was
leaving it behind. He could not have passed the moon
thus. He went towards the light, and when he had
reached it, it seemed like a slender staff of light. He

touched it with his hand, and although he did not feel anything, yet he could take hold of it, and he walked on with the slender beam in his hand.

He had not gone very far when in his walk he touched on something lying in the path. Bending down and touching it with his hand he found that it was the form of a fellow creature. "He is overcome with fatigue; can I help him along?" he thought. He rose up to look round, and let the beam of light which he held in his hand touch the prostrate form. "I wish he could get up by himself," he thought. No sooner had he felt this wish than he had a sensation of pain in his limbs, and the figure rose up.

"I could not move," it said, "until you came, with all my reasons to get along; the pain was as much as the pleasure."

"Who are you?"

"I am a wanderer, and am trying to reach the place where I was born; they will help me there."

Now in the valley there was a certain set of people called wanderers, who had proved themselves unfit for any real work. These, if inoffensive, were allowed to roam about subsisting on charity. The student walked alongside this wanderer; and every step the wanderer made he felt a sensation of pain in his limbs. But the two walked quickly on till they came to the dwelling he had left so shortly before. The student led him in and let him rest in his chamber. And then he himself left the dwelling again, taking with him a few necessaries.

CHAPTER IX.

WHEN he had seen the wanderer safely housed he determined to go and visit a friend who had lived in a town not very far from the metropolis. This friend had been his most intimate companion when he first became a student, but being older had finished his studies sooner, and had left the metropolis before the student's misfortune. In leaving his place of exile the student rendered himself liable to punishment, and he gave up the means of subsistence which had been provided for him there. He was obliged to go as a wanderer, and trust to the liberality of the people on the way.

He was hospitably received as a rule. The region was remote from the metropolis, the inhabitants were glad to talk with a stranger—and the wanderers were, in general, held to have a stock of exchangeable talk and news. But he did not speak with any one of what lay present to his mind, till one occasion.

As he was walking along early in the day, he was hailed by an inhabitant who looked like a well-to-do farmer. Something in the student's appearance attracted him, for, learning that he was on his way to a distant town, he asked him to stay and take the first meal of the day with him. This inhabitant had been a clerk employed in the council of pleasure and pain. But the sedentary life had been too trying for him ; he had come to live in the country on a small possession of his till he had overcome the strain.

"Did you not find it very dull in the part you come from ?"

"No ; I found that the people had much of interest to tell me."

"They have singular traditions. I remember when

a deliberation was held in our council as to whether they were pernicious or harmless; it was decided that they were harmless and little likely to spread."

"I have talked a good deal with them since I have lived amongst them, and have come to the conclusion that in what they believe a great deal is true."

"Indeed! you cannot surely believe that our pleasure is distasteful to any being outside us."

"No; but I go back to the old notion of which you have heard, that there is a being who calls us into being, and who is over us; and I believe that this being takes pain, and so makes life pleasurable to us. You know that some sensation is passing away, and you know that there must be more pain that passes away than pleasure."

"How can I know that?"

"We know that there is not such a very great excess of pleasure over pain. Now if in all the course of time that has been, the sensation that has been passing away was pleasure, there would by this time have been left an excess of pain, and before now we should all have sunk into apathy. So it is either pleasure and pain mixed which passes away, or pain alone. I conceive that it is pain alone. These strange doctrines are true, only curiously expressed. The being over us is continually bearing pain and so making existence pleasant to us, thus causing us to move and live. So the pain of our life is that remaining pain which he does not take."

"This seems to me a very dismal doctrine. I can imagine some poetry in the idea of a being of infinite power, strong and glorious, but none in the idea of a suffering being."

"When you were a child you thought your father could do everything; but as you grew up and found that he too had his difficulties, was your regard for him lessened, or your thankfulness for that which he did for you?"

"No. And you mean that if we do not regard this

being in the same way, granting his existence, still we should feel gratitude towards him."

"Certainly we should feel gratitude to him ; and, considering the attitude we have taken towards him, this feeling of gratitude comes over us with a kind of revulsion. But besides gratitude I do not see why we must lose any other feeling such as you seem to miss. Do you not remember how, in the course of the studies we have all been through, we were told that there were two parts in knowledge—one corresponding to reality, one introduced by the action of our own minds—so that certain characteristics which we at first think to be due to the nature of things in themselves we find out on reflection are only our apprehension of our own mental action ? "

"Yes ; we do not perceive the reality absolutely, we apprehend it subject to the mind's mode of perceiving."

"And of course the mode of the mind's action makes it perceive certain qualities as parts of the real existence, which do not belong to real existence at all. These qualities spring from our mind's own action. In old times these qualities were considered to be qualities of the reality instead of introduced there. And much of the impressiveness of the idea formed of the being of whom we speak was due to a mere magnification and extension of these qualities—qualities which do not correspond to anything in reality. So the impressiveness of the idea of this being was due to the magnification of qualities which originate solely in our minds."

"This accounts for the idea having faded away. But tell me definitely in an instance. Explain by taking some particular quality what you mean."

"I cannot do that, the thought but floats in my mind ; still it is always good to embody. Something of this sort. When we observe any object we always attribute to it a certain power. Everything has its own powers of resistence, of moving, of affecting us in certain ways. Thus whatever we apprehend, we apprehend as powerful. Now since this quality of powerful comes in with regard to everything, it is probably introduced by the mind, and is rather a part of the mental action in giving an idea of reality than a quality of reality. If so, when we suppose a being to have the quality of 'all powerful,' we are not supposing anything at all about the being, but are only extending a quality quite barren of any correspondence with the absolute nature of things. We have left off talking about the being, and are extending a conception which springs solely from the only way in which we can perceive."

" Surely you would say that this being was powerful."

" Of course, if we think of him at all, we must conceive of him as powerful ; the nature of our mental action demands this. But to dwell on the notion of his powerfulness is quite barren, the only subject of thought which has content is to inquire what kind of power he has. There has been a tendency on the part of those who have thought about this being to represent his greatness in every respect. But they have not always been judicious in so doing, because being unable to separate his real qualities from those which they attribute to him in virtue of their own mode of perception, they have come to lay stress on descriptions which on the one hand correspond to nothing in reality, and on the other hand fail to move those whom they are intended to impress. A cloak has been woven. The nature of this being is hidden. His nature has been connected with introspective questions about the origin—of, of all things, the way in which we perceive. All this must be dashed aside. This being is the cause of all our life, and yet he needs your help as you understand help."

"I should like to accompany you to your friend and hear what he has to say."

"Come, certainly."

So they went together to the town. On the way the clerk felt a brightness of existence such as he had not enjoyed for a long time. They talked together, and confided in one another. At length they came near the town where the student's friend lived. They separated, the clerk going into the town, the student to the house of his friend. On his way there the path led through a small wood of very thick growth. Passing along, he found that he had left the path. Pausing to reflect in which direction he ought to go, he thought he heard a sound. It was repeated. Penetrating deep into the obscurest part of the wood, he searched till at length he found—carefully concealed—a child, a mere infant.

The child was nearly perished with exposure. He took it up and warmed it. When the child was a little better the cause of its having been hidden away was apparent. Its breathing was distressed and laboured. It suffered under some affection of the lungs, which made it gasp at every breath. Still in other respects the child was well developed and seemed strongly made. It seemed to have been left too long without care to recover. The pain of exhaustion from the neglect, and added to this the pain of its breathing, was too much for it, it was sinking.

"If I could bear the pain of its breathing," thought the student, "it might not sink till I could get some nourishment for it."

He looked up, for it seemed to him as if some one struck him in the chest. There was no one there. The pain continued. He did not drop the child but continued on his way to the house of his friend. When he got there he noticed a stillness unusual in the houses of the inhabitants. He entered, and was met by his friend's sister. He saw at once that something must have happened. She took him into a dimly lighted room, where he saw his friend lying motionless and his face quite white.

"He has been suffering great pain for long," she said; "it was hoped that if he could bear up the pain would have run its course and he would not sink. But all we could do was no use." The room was full of all things accounted pleasurable, and she looked round as she spoke. "It was no good." Taking the child from his arms she left him with the form of her brother.

Sitting down by his side the student felt the strange oppression on his chest continue. He went out and found that the child had completely revived. It had still the appearance of being agonized in its breathing, but its eyes were bright, and it laughed.

"It will be all right soon," said his friend's sister.

"Tell me what was the matter with your brother."

When he had heard about his malady he returned to the room. After he had sat there for some time he felt more and more the sorrow for the loss of his friend, and the need of his counsel. This aimless, inert form, this lifeless mass, was that which he had come to seek—was the being with whom he had longed to confer.

He bent over him. "Could I but snatch him back into life; could I but have one hour's intercourse with him. If I had been with him I might have borne some of the pain of his complaint before he was overpowered with it." He touched the lifeless hands, they were cold and damp. He gazed into the expressionless face. He seemed to feel the pain of the inner struggle his friend had waged against the disease. The quiet of that still chamber was gone for him; in his own person he felt the pangs of the struggle for life. A mist came over his eyes, and he sank down holding his friend's hands. Suddenly he heard a voice. He rose and looked about him. The sound came faintly from the lips of his friend.

"I have been very ill," were the words he caught. "I am so glad you have come; I was thinking of you in my worst moments. You have come just as I am getting better."

Indeed the features were regaining expression, the hands were warm. It was his living friend again.

After a few hours he was sufficiently recovered to hear about all that had happened. They talked together long and earnestly. His friend was convinced.

"Let us go to your companion," he said.

They went into the town together. They found that the clerk had gone to the magistrates' hall where a trial was being held. They did not see the clerk at first, so they listened to the proceedings. A woman was brought in who had been kept in prison for some days, accused of concealing her child. The case was clearly proved. The woman received her sentence with an appearance of apathy.

"She will not come out of prison alive," said the student's friend, noting her expression.

But he called out to her from where they stood in the body of the court, "Do not fear, your child is safe."

The woman's face brightened, and she went with her jailers buoyantly.

The magistrate had remarked who it was that had spoken, and was about to give orders for the disturber of order to be brought up for punishment. But the clerk, who was sitting near to the magistrate with whom he was acquainted, said:

"This is the one I have told you about; pray do not punish him."

The magistrate accordingly contented himself with warning the audience in general terms.

But he said to the clerk, "Something about him is very repulsive to me, do not tell me anything more about him."

The three returned together, and together they deliberated as to how the new idea about the king could be made known. It seemed best to go to the metropolis and talk with the wisest and most learned there.

The student asked about the child. His friend's sister came and told him that its breathing was not any better, but that the child itself was strong and playful.

"It belongs to the woman who was tried to day," said the student, "and must be kept safely till she is out of prison."

His friend after some deliberation gave it in charge to a faithful servant to take to the metropolis. A suffering child there would be much more likely to be overlooked, "and you," he said, "will be able to look after it."

As the student and the clerk were about to set out on their way to the metropolis his friend took him apart.

"My sister tells me that I had sunk into apathy when you came."

"Yes."

"And that you called me back?"

"Yes."

"How can I thank you! had it not been for you I should never have enjoyed life again. I am grateful to you."

"Do not say grateful to me, but rather to that power which does for you all your life that which I do for you momentarily now. And even now it is not to me that you should be grateful, but to him, for it is only because he has enabled me to do so that I have taken of your pain."

With this he took farewell of his friend, and with the clerk proceeded on their way.

They had not got very far when a train of servants came up behind them. They stood by on one side, but from the midst of his attendants a youth stepped forward.

"I have learned what you have done, and I have overtaken you with great haste."

"What is your wish?"

"I want to come with you. I know that you have restored your friend from apathy to life. No power is so great as that. I have riches in abundance. All that I have is at your service ; teach me your power."

Now in the valley riches meant abundance of pleasant things. At the time the student was bearing the constant pain which he took of the child's breathing, and the pain also of his friend's illness. He felt that before beginning to take pleasure—which was the meaning of having pleasant things—it would be necessary to give up the power which he was exercising, so he said to the youth somewhat harshly :

"You cannot compare riches and that which I do, nor can you exchange the one for the other. First give up all your riches, then you can begin to learn what I do."

The youth turned back, but once again spoke, saying :

"I will give up a great part of my riches if you will teach me."

"If you want to keep any, however small a portion, you cannot do what I do."

Then the youth with all his attendants passed away.

CHAPTER X

WHEN they came to the metropolis the clerk brought many of his acquaintances to see the student. From his position in the council chamber, he was able to address and induce many of the ablest of the councillors to come and inquire. But as soon as they came into the presence of the student a sort of constraint sprang up between them. They did not take his words as having any real meaning. They were occupied all the time on speculating what motive it was that made him say these things, and as to what kind of difference it was which they felt existing between him and them.

In fact, as time passed on, no one of any position or power would be brought into any sort of approximation to him. On the other hand he used to speak continually with the poorer people. Those that were sick especially delighted in his presence. There seemed to be in him a power of stimulating those that were sinking into apathy back again into life. Those who were worst off in the city seemed to feel when he spoke to them a promise of an alleviation of their sufferings.

One day the clerk asked him

" How is the child ? "

" It is well."

" But it still seems to breathe with as much difficulty."

" Yes, but see how happily it runs about."

" How do you manage to preserve it? Any child which I have seen would be pining miserably with such an affliction. What is the power which the being you tell of has given you ? "

" It is no power in the sense you mean."

" Surely it must be. Have I not followed you faithfully and done all I could to get the wisest in the city to listen to you ? Surely the time has now come when you will tell me what this power is, and, if you can, let me share it."

" You do not know what you ask."

" Tell me, I pray."

"It is simply this, when I became aware through thought of the being that is over us I had no message or command from him. But I found that I could when I stood by any suffering being take some of the suffering and bear it myself. So as he of whom I tell does with us each moment of our lives I do occasionally and in a little manner."

"But what pleasure do you get that makes all this worth your while ? "

" There is no pleasure. I am glad to see the being freed from suffering, and living instead of sinking."

" Do you mean to say that there is nothing to hope for ? "

" I hope the time will come when I shall have a fuller knowledge of the being I know."

The clerk was silent. He went out. While he was still thinking over what he had heard in answer to his inquiries, a messenger came to him from the chief of the councillors of pleasure and pain, asking him to an interview.

When the clerk had been ushered into the presence of the chief councillor, and was alone with him, the latter said :

" I should like a little quiet conversation with you about your companion."

" I shall be glad."

" When you gave up your office and retired you had no expectation of being concerned in affairs of state again so soon."

"I did not expect, certainly, and I do not know what your meaning may be about my being concerned in affairs of state."

"What I mean is very simple. The continued deliberations, generation after generation, of the wise men who assemble in the council chamber have been the cause of the continued progress of the inhabitants. Nothing is done by them hurriedly or violently, but gradually improvement after improvement is worked out. But besides this, there have always been at every age certain disturbances in the state ; certain doctrines are brought forward, and sometimes these tend to good, and should be encouraged ; sometimes they are of unknown import, and must be studied ; sometimes they are against the happiness of the state, and then the grave responsibility rests upon us of checking them. Now from your position you have more opportunity of knowing than any one else in what direction your companion's doctrines tend. I have sent for you to ask you to share with me this grave responsibility."

"I do not think I can help you. I am sure he does not wish to do any harm. What harm can there be in his doctrines ?"

"It is not so much about his doctrines which I want to speak to you as about another subject. Many of those who have talked with him have agreed with one another in ascribing a singular oppressiveness to his presence. The expression was even used by a very worthy friend of mine, 'He made me feel like a puppet.' Now what right had he to inflict such a sensation on a very worthy individual ? I want to ask you yourself if you have ever felt this ?"

The clerk hesitated.

"At least, tell me, have you ever found it easy to influence him ?"

" No ; I do not feel as if I could influence him in the least. He seems to lack the ordinary springs of motive."

" Now, should you say that it would be a gain to the community if many should become like him ? Would not they be difficult to govern ? "

" Certainly they would be difficult to govern."

" Would it be a gain in pleasure to the rest of the inhabitants or to themselves ? "

" It would not be a gain to themselves," said the clerk, recalling the pain which his companion bore, " but it might be good for the rest of the inhabitants."

" Yes," said the chief councillor, " that is where his strength lies ; he is a very skilful physician or an impostor, and he has the people on his side from the cures he has effected. Can you tell me anything about his life ? "

" I have heard from him that he was a student, and was exiled ; and that in his place of exile he found out the new doctrines, and he left the place he was sentenced to. On his way I joined him."

" So much we know, and it is within our power, according to the regulations, to compel him to go back, and to punish him for having left the region he was banished to."

" If you have that power, why do you not send him back if you think it would be best for the state for him to disappear ? "

" Ah, my good friend, you have heard a great deal of our public deliberations from your place in the council ; but now that we are consulting together, I must tell you that there are deeper secrets in the art of government, which you will readily apprehend. Suppose we arrested this individual and sent him away, the people would not see the justice of it. They want him now, and they would say that the forms of law were being used to get rid of him. Of course if his partizans became violent something of this kind would have to be done. But it is only a decree that seems just in the eyes of the people that we can prudently carry out in such a case without attracting even more attention to him than there is at present."

The clerk said nothing. The chief councillor went on :

" I am sorry that our conference has come to so little. I was hoping that I might have found in you a successor to the vacant seat in the chamber. I know you have the ability to fill it well. But before the advancements are made some proof of the wisdom of the successor is required. Hitherto you have not had the chance, but I thought that in this difficult case, where you have so much better opportunities of observation than any one else has, you might have shown your mental power and confirmed my opinion of you. Still, no doubt, on some future occasion you will have another opportunity when this affair, difficult as it is, is forgotten."

The chief councillor made a sign that the interview was at an end, but the clerk remained.

" All that we want," the chief councillor resumed, " is to form an opinion from inside knowledge of whether this innovator is likely to cause more pain or more pleasure if he gains a hearing. Can you advise us ? any particle of knowledge of his inner life, apart from his public professions, is valuable."

" There is a singular fact which I should like to tell you of, as it has been somewhat of a burden to me."

The chief councillor made a sign of assent, and the clerk told him about the child, and how it had been preserved.

" And with this child," he said, "he and I sit when the day's work is done."

" It is indeed a strange story," said the chief councillor ;

"you are quite right in telling me. I was sure you were one on whose discretion confidence might be placed. You have given me the highest proof I could have expected. The bearings of this matter must be thought over."

That evening, as the clerk entered the room where they lived, the student was leaning over the child with a wearied expression. He went up to him and laid his hand on his shoulder. The child looked up at them and laughed. It was quite happy despite the apparent struggles of its breathing. The student looked at his companion's face. His weariness vanished at once, and a strong warm light came into his eyes.

"You seem oppressed, my friend. I know you regret the way in which all the wise and important people you have brought here look on me, and you must feel some sorrow for the partial loss of esteem they have showed you in consequence. Can I help you to bear it?"

At that moment the door opened, and a messenger came in and gave the clerk a sealed packet. He opened it and saw that it was his appointment to the vacant seat in the council chamber. But his face did not brighten. He answered his companion moodily, and thus the day ended.

CHAPTER XI.

ON the next day the student rose early and went forth alone. He did not, as was his wont, go amongst the people, but he passed through the streets towards the open country. On his way he was stopped by an old woman, bent with age and many infirmities. She had no place amongst the people, and had so many pains and such a barrenness of existence that any one who had thought of her would have wondered that she remained alive.

She stopped him and said, " Master, I have heard that you can take my pain. Help me."

But he answered, looking at her, " No, I cannot, but I have a message for you."

And she said, " A message for me? I do not know any one who would send me a message."

But he answered, " Nevertheless, I have a message to you from my lord, and he bids me thank you."

She answered, " It cannot be. You must have made a mistake."

But he said, " I have made no mistake ; he thanks you."

He could not explain to her how by her bearing pain, according to the law of the valley, she took it from that which the king bore. Instead of saying that, he gave her the message, and somehow the old woman believed it.

The rest of the day he spent in the open country. When he returned it was getting towards dusk. There was an unusual movement in the streets. On passing into the public market-place he saw a crowd collected ; and when he had penetrated to their midst, he saw lying on the ground the child he had kept so long. It had been lying uncared for and exposed for many hours ; and the want of food, the fright, and its gasping breathing made it the most pitiable object. He at once stepped towards it and took it up in his arms.

" Is that your child ? " said one of the crowd.

" It is not my own," he answered, " but I take care of it."

" Then it is you that are bringing pain upon us all," shouted several voices from the back of the crowd. And some one shouted out :

"I know you. You pretend to take pain away and you really bring much more in secret."

And moved with a feeling of indignation against the one who had caused such a painful object to exist as the child was, the crowd closed on him, and barred his way to his own place. But they did not lay hands on him. As he stood with the child it gradually began to regain its composure. But with a sudden movement the crowd swept towards the council chamber. And when they had come there they demanded that this cruel and wicked act of keeping pain in existence should be punished.

There happened to be several of the chief magistrates on the spot, and in obedience to the voices of the crowd they proceeded to sit in judgment at once. It was not known how the child had come into the streets ; but it was admitted by the prisoner to be his doing that it had been kept alive. The doctors unanimously said that it ought to have been put out of existence directly it was born. There was practically no defence. The charge of subverting the laws was established. The people clamoured for the extreme penalty. The judges passed sentence on the student.

Before morning he was put to death.

He met his fate without sorrow, even with gladness. The pain in his life had for long been as much as he could bear. He did not, like the prince of long ago, look upon nothingness as the desired end of existence. He felt the presence of the one whom he had discerned through thought, and this seemed more real to him than life or death.

On the following day, whether in reaction from the excitement of the previous evening, or from some other cause, an unusual quiet pervaded the streets of the city. There was not much discussion as to the event which had happened. The prevailing feeling was one of wonder that there should have been so much commotion about an unimportant affair. For the most part before the next evening the whole circumstances were on the way to be forgotten. And yet every here and there were persons in whose lives the loss of their friend was deeply felt. The joy and spring of life seemed gone. The poor child lay pale and motionless, save when every now and then it gasped convulsively for breath. None felt the despondency more than the clerk. The interest and value of life seemed to have gone. He did not care for his new honours.

That day some most unexpected news went through the town. The chief of the council of sensation had sunk into apathy. He was in the prime of his life. It was most unexpected. Every one was astonished at the news, but were still more astonished at how little they felt concerned.

Following on these tidings came others. Many of the inhabitants of the metropolis whose lives were most strenuous suddenly succumbed. The clerk had made up his mind to go into the country. But tidings came from there also that the poorer labourers, and those who were exposed to the fatigue of long journeys or exposure were in many cases sinking. The wave of torpor seemed passing over the whole valley and not to be confined to the metropolis. The rich and unoccupied classes only were comparatively unaffected. They betook themselves to the store of enjoyable things at their service, and so replaced the natural spring of life which seemed tending to fail in every one.

On the confines of the valley, where the ravine struck it vast depth between this land and that, vast and endless as the sea stretched the plain whence the king had come. It was struck silvery grey by the light of the moon, dark shadows marked the nearer strands, and gradually the rocks which cast them showed their sharp outlines, hardly distinguishable from the ground out of which they rose.

Over the great gulf floated the sounds of a pipe, the strains were low, winning the soul with the sweetness of an unearthly melody, throbbing as with a call to a distant land away and beyond.

And when the eye found the source of the sounds, there stood, once more, solitary in the untenanted vast, the king's devoted friend, the same old man who before had hailed him. Gradually the music sank lower and lower, till at length silence spread in folds unruffled. Then on the edge of the valley a form appeared. It came and seemed to gaze across the gulf, standing motionless and intent. At length a voice came.

"Art thou there?"

"Yea, O king, what wouldst thou? Art weary?"

No answer came.

Then the old man spoke. "Behold the roads where they stretch gleaming white in the moonlight; behold the fields, the villages; see in the distance the great walls of the palace. Have not these risen up for thee, O king."

Then the king made answer: "I am weary."

Suddenly the old man raised his pipe with both his hands to his lips. Wave after wave of triumphant sound pealed forth. Great harmonies such as marching nations might hear and rejoice, noble notes of unbounded gladness.

Then, crossing by an unknown way, he came and stood by the king's side. After a while the two moved on together, and by a secret path passed away from the valley—whither I know not.

As soon as the king had departed from the valley the

beings in it began to sink into the same state of apathy
as those were whom he had first found there. Those
who sank first were the ones in whose lives the stress of
labour or thought was the most intense, for they first
felt the loss of that bearing of pain by one beyond
themselves which gave them a difference of pleasure.
And slowly as the accumulated enjoyment was exhausted, a chill death in life crept over the land. 'Tis
useless to ask after the fate of any one of those that
were there, for each was involved in the same calamity
that overwhelmed all. Every hand forgot its cunning.
The busy hum of life in the streets was hushed. In the
country the slowly moving forms gradually sank to rest.
At every spot was such unbroken quiet as might have
been had all the inhabitants gone to some great festival.
But there was no return of life. No watchful eye, no
ready hand was there to stay the slight but constant
inroads of ruin and decay. The roads became choked
with grass, the earth encroached on the buildings, till in
the slow consuming course of time all was buried—
houses, fields, and cities vanished, till at length no trace
was left of aught that had been there.

PART II.

CHAPTER I.

THERE are certain respects in which our world resembles
the valley. Instead of regarding pleasure, pain, and
feeling, let us examine the world we live in with regard
to motion in one direction and another, and in respect
of energy.

If we observe the movements which go on in the
world, we find that in great measure they consist of
movements which if put together would neutralize each
other.

A pendulum swings to and fro. If the two movements took place at the same time the pendulum would
be still. Taking a more ample motion—that of the
earth round the sun. The earth moves in the course of
its orbit as much towards the sun as away from it, and
as much towards the east as towards the west. If all
the motion were to be gone through at one and the
same time the earth would not move with regard to
the sun.

Again, if we notice what goes on on the surface of the
earth, we see that there is a motion of rising up and
of sinking down. There is an approximation of the
chemical elements into some compounds, and a separation of them again. Of all the myriad processes which

go on, the swing of a pendulum is the type. But the downward swing may be very different to the upward swing. It may be that the downward swing is represented by the violent action of the chemical affinities in a charge of gunpowder when exploded, and the upward swing may be represented by the swift motion imparted to a cannon ball, and the swift motion of the cannon ball in its turn comes to rest, and as it comes to rest slowly or quickly other changes take place.

And what we notice in our world is similar to what the inhabitants of the valley noticed about pleasure and pain—that they do not neutralize one another as a matter of fact.

The contrary motions on the earth which, if they were put together, would neutralize one another, do not as a matter of fact neutralize one another. We call motion in one direction positive — in the opposite direction negative. But in the world as a matter of fact positive and negative motion do not together come to nothing.

As in the valley the states of pleasure and of pain did not coalesce into a state of apathy, but always succeeded one another, in simple or complicated fashion, so on the earth it is impossible from two opposite moving bodies to get stillness. If the two come into contact in opposite directions the movement does not stop, but makes its appearance in an alteration of the shape of the bodies, in a disturbance of their particles, or in some such fashion.

Again in the valley, by measuring the pleasure and pain simply as feeling, and not taking into account whether it was pleasure or pain, the inhabitants found that the feeling was always the same in amount.

So we on the earth, measuring the amount of movement, and leaving out of account whether it is positive or negative, come to the conclusion that the quantity of

movement, reckoned in the way in which we call it
energy, is always the same. The principle of the conservation of energy has become a fundamental one in
science.

But besides the discovery that the amount of sensation
as such was always constant, the inhabitants of the
valley discovered that a portion of the sensation was
passing away from a form in which they could feel it.

And there is an analogous discovery in science. We
know that a portion of the energy of our system is
passing away. It is not being annihilated, but is disappearing. With the energy which can be collected
from the falling of a stone, the same stone cannot be
raised to its former level again. Some of the energy
has disappeared from the form in which it can be known
as the energy of moving masses. The energy has in
some measure irrecoverably passed off in the form of heat.

Hence, just as the inhabitants of the valley came to
the conclusion that in point of sensation they were
"running down," and that after a time all sensation
would have passed away from the form in which they
could feel it, so we have come to the conclusion that
the energy of the system in which we live is running
down, that the energy is passing out of the form in
which it can be manifested as moving masses, that
finally all movement of masses will come to a standstill,
and there be nothing left save motionless matter, with
warmth equally diffused through it.

Now in coming to the conclusion about the valley,
that the amount of sensation was gradually passing away,
the inhabitants, as we have seen, had come upon the
very secret and cause of all the life in the valley. But
coming upon it from the outside they had not recognized
the significance of what they had found. The cause
and prime mover of all their existence indicated itself to

them, coming thus upon it, as a process whereby all that went on was doomed to a distant but certain extinction.

Now, is this process of the passing of mechanical energy into the form of heat to be interpreted by us in a way analogous to that in which the inhabitants of the valley could have interpreted the process they found?

In this cessation of sensation in the form in which they could experience it lay the central fact of the life of the valley. Has this passing away of energy from the form in which we can experience it an analogous signifi-cance to us?

In order to examine into the possibility thus suggested there are four convergent lines of thought which it will be well to follow up separately. Each of these lines of thought bears in an independent manner on the central question—the significance of the passing away of energy. These lines of thought may be connected with the following words, which indicate their significance: (1) Permission ; (2) Causation ; (3) Conservation of Energy ; (4) Level.

CHAPTER II.

WHEN we observe any movement taking place we ask what is the cause of it? what is the force which produces it? But surely, if we confine our inquiry to this point, we have made an omission. That we are not conscious of having made an omission may perhaps come from our living in the air which yields so easily to any moving body. If we lived in a rigid medium we should, when we became aware of any moving body, ask two questions. First, what urges it along? secondly, what prepared the channel for its motion?

But seriously, without laying any stress on the above

illustration, we see that to every movement two conditions are necessary: a pushing and a yielding, a force
and a permission. If the particles of the air could not
yield, a pendulum could not swing through it. If again
the air could not pass on the motion it has received, it
could not yield to the motion of the pendulum.

Now since every motion requires a permission, we are
led to ask the question, What is the ultimate permission? What again is that which by yielding allows
motion at all to take place?

If we trace any movement scientifically we find an
indication of what the ultimate permission is.

A body swings through the air. Currents in the air
are set up. These currents impinge on the objects with
which the air is in contact, and in them produce heat—
producing heat also by friction with other portions of
the same air. Every motion thus passes off finally, at
however long an interval, in the form of heat. Motion
may reappear as motion through myriads of phases, but
at each change of form some of it passes off into the
form of heat, and finally all passes off into the form of
heat. Thus, unless matter admitted of being warmed,
there would be no ultimate permission. A motion once
started would never come to rest. Or, rather, no motion
could take place at all.

The tendency of the above remarks is to avoid the
conception of there being absolute laws of motion, true
of bodies when surrounded by no medium, modified
when a medium is present. Surely such a conception is
an instrument of the mind for exploring nature, not an
absolute fact in nature. The abstract laws of motion
are mental aids in creating knowledge; like scaffolding
for the builder, even from their very usefulness they
have probably but little to do with the permanent
edifice.

This passing into the form of heat supplies a place analogous to that of the "void" in the speculations of the Epicurean philosophers. They argued that motion was not possible without a void. Given a void, somewhere into which matter could move, then any amount of motion could be accounted for. But without a void into which a portion of matter could move, how was it possible for motion to begin?

Thus repeating their inquiry with our altered conceptions, we ask this question about motion, or energy (which is a particular way of reckoning motion).

Unless motion can in some way pass off, how can there be all these transformations of energy?

Now the ultimate transformation of all energy of motion is into the form of heat. In this change into the form of heat is to be sought the ultimate permission which makes all transformations of energy, all motions, possible. It is this being acted on of the finer particles of matter which permits the movements of the larger masses.

This passing of energy into the form of heat must not be regarded as a side circumstance, as less essential to the laws of nature than that law which we call the conservation of energy. It is at the same time the end of every motion, and that which makes every motion possible.

The passing of energy into the form of heat takes place in that which we call friction, and in all those modes in which any movement is brought to a standstill. But so far from these being simply "hindrances" to motion, it is through them that we learn that which makes motion possible. It is with us as with the inhabitants of the valley, the gradual cessation of feeling from their life and the modes in which it ceased were the way in which they regarded the action of the king who was the cause of all.   We have thought of motion as a thing in itself impaired by the multitudinous obstacles it meets in the world.   Let us look on the circumstances more impartially. Let us look on them as something co-equal with motion.   Let us find in that mode whereby all motion comes to an end the originating cause also whereby all motion comes to be.

The passing of the motion of masses into the form of heat is the ultimate permission.

CHAPTER III.

IF we reflect cautiously on the history of our opinions, we find that we often fall into error in respect to our freedom in attributing causes.   If we are unfortunate we are apt to look on our neighbours, or the world, or, if we are of a self-depreciatory turn of mind, ourselves as the cause.

Again in past times people really felt sure about certain things being causes which we now know had a very slight connection with the result.   Incantations have been supposed to have an effect on physical phenomena, such as eclipses.   Numbers and their properties have really been conceived as the causes of the modes of existence.   Ideas have been supposed to have causative power over the order of the world.

We should be very careful in attributing the notion of causation.   If we see a stone lying on the ground, and proceed to pick it up by the strength of the arm, we say that the exertion of the arm is the cause of the stone being lifted.   But in this respect even we are too hasty. The arm may exert itself and yet the stone not be lifted up—if it is too heavy.   All that we can say about it is that if the stone is lifted, a certain set of muscular actions has gone on in the arm, and a certain movement of the stone has taken place. If we look closely at the matter, the movements in the arm are related to the movements in the stone in a strictly measurable way. There has been so much exertion corresponding to the weight of the stone. But suppose the arm had done anything else, there would have been the same relation traceable between the movements in the arm and the actions which followed its movements. The energy spent by the arm would be equal to the energy imparted to the object moved, whether it be a stone sent flying through the air, or one lifted to a higher position (bearing in mind always the small quantity of energy passing off in the form of heat).

It does not seem advisable that the notion of cause should be brought in to denote the relation of the movement of the arm and the movement of the stone. These are two sets of actions between which the regular relations which hold good between the consecutive states of moving systems hold good.

The notion of "cause" should rather be applied to that act of the will whereby the movements of the arm are connected with the movement of that particular stone rather than the movement of any other object.

We are the cause of the actions we will. The notion of cause is derived from our "will" action, and the notion of cause ought to be kept to this connection.

All that goes on outside us can only be apprehended as consecutive states following on one another. Between certain sets of consecutive events we notice that the same relation holds good which we have observed in other consecutive states. If some water is heated in England it passes off into steam ; if water is heated in another part part of the world it also passes off into steam. There is an exact analogy in the behaviour of water under the action of heat wherever we observe it. But all that we have obtained as knowledge is the fact that we may practically be confident of an analogous behaviour on the part of water wherever circumstances are similar. We may use the expression that heat is the cause of water boiling for convenience. But the expression should not be used as containing any deep meaning. To say one external event is the cause of another is to put an absolutely unknown and spiritual relation in place of impartial observation.

To cause a motion is the name for the action of our soul on matter—a thing shrouded in mystery. To be the antecedent in a chain of movements is the fact which we can observe about any movement in the external world. We cannot strictly say what movements of gases, water, &c., cause this volcano. We can only say what movements of gases, water, &c., precede this volcanic eruption analogous to movements which have preceded other volcanoes.

There are invariable sequences in the external world to which we do not affix the notion of cause and effect—day and night, summer and winter. Why we should do so in any case is not clear, except that by familiarity and mystery the sequences have become to us something like our own will action. Indeed, is it not the case that when we can trace intermediate links we say so and so comes from so and so in such a manner. But when no intermediate links can be traced we say one event causes another.

If, however, we omit the feeling of causation from the external chain of events, it does not follow that there is no causation to be apprehended in the external world.

Let us not introduce the notion of causation at haphazard. But if we find in the external world signs of an action like our own will action, let us then say, Here is causation.

The inhabitants of the valley would not have been right in saying that one act of a routine caused another: But they were right in saying that the amount of sensation was constant, and that some of it passed off in a form in which they could not feel it.

And so let us not say that one action causes another. Let us not say, for example, that the downward swing of a pendulum is the cause of its upward swing. But let us simply say that the one follows the other ; that the amount of energy present is the same except for the small portion that passes off into the form of heat.

CHAPTER IV.

SUPPOSE certain sets of numbers were being presented to us one after the other, and amongst these three consecutive sets were the following. First set: 3, 5, 6. Second set : 8, − 2, − 1, 1.   Third set : 7, 4, 2, − 1.

A little consideration will show us that there is a certain uniformity in these sets.

Take the square of each of the numbers in the first set and add them together, the result is 70. Thus $3^2 + 5^2 + 6^2 = 9 + 25 + 36 = 70.$

The sums of the squares of the numbers in the second set come to the same. $8^2 + (− 2)^2 + (− 1)^2 + 1^2 = 64 + 4 + 1 + 1 = 70.$ Also in the third, $7^2 + 4^2 + 2^2 + (− 1)^2 = 49 + 16 + 4 + 1 = 70$, and so on.

Having noticed this we should regard it as a purely formal law, having nothing to do with why the numbers were presented to us. But we should consider it likely that it would characterize all the numbers that were presented to us. And if this expectation were found to be realized, we should after a time feel a certain assurance that the next set of numbers presented would satisfy the same law. If this assurance was indefinitely satisfied we should get to regard the satisfying this law as an invariable condition of the numbers presented. But we should never regard this purely formal law—that is, a law about the particular characteristics of the numbers—we should never regard this formal law as the cause of the next set of numbers appearing after the first had gone.

When, however, we talk about the conservation of energy we are apt to think of it as more than a merely formal law, more than a statement about numbers which has been found to hold true.

Yet it is no more. The law of the conservation of energy asserts that in any system in motion the sum of the squares of the velocities of the particles at any one moment is equal to the sum of the squares of the velocities of the particles at the next moment.

The conservation of energy is but a mode of reckoning motion, by which it is found to be constant in all changes of a system. The system must embrace all the particles concerned in the motion. It may be made as large as we like.

The principle of the conservation of energy as here stated is confined to the case of moving bodies. Sometimes the energy is said to disappear from the form of motion and become potential energy. That case will be treated under the fourth consideration of level, but it introduces no alteration in what has been said.

As to the practical truth of the law of conservation of energy there can be no doubt ; nor as to the value of the results obtained from tracing its validity in obscure actions. But there is nothing final about it. It is a numerical statement of extreme value, and it introduces a mode of reckoning by which motion can be looked upon as indestructible as matter is.

There is a possible objection to the law of conservation of energy.

It is no less a law in nature that in every one of a series of changes some of the energy passes off into the form of heat. Now heat is reckoned as a mode of energy. And there is in science a method of calculating how much energy any given quantity of heat is the equivalent of. And this equivalence is calculated on the supposition that no energy is lost. When heat is produced and motion passes away, the proportion between the motion that disappears and the heat that appears is represented by a number calculated on the assumption that no energy is lost. Thus whenever any quantity of energy takes the form of heat, the quantity of heat which is produced is exactly given by the calculation. But the reverse process is not possible. It is not possible to turn back all the energy in the form of heat into the form of motion. Consequently it cannot be proved that the energy in the form of heat would, if all turned into motion, produce as much motion as that from which it was produced. There may be an absolute loss of energy—only a very small one. The law of the conservation of energy may be the expression that this loss is a minimum.

This objection is not essential to the line of argument pursued above with regard to the conservation of energy. It forms no necessary part of the line of thought we are pursuing. It merely tends to show that the law of the conservation of energy is no axiom which we cannot suppose not true. The real conclusion to which this part of our line of thought tends is that the conservation of energy is a purely formal law.

8

CHAPTER V.

THE most apparently simple movements are those which we see taking place on the surface of the earth, connected with the agency which we call gravitation. We see the rivers flowing from a higher to a lower level, rocks when loosened from a mountain side rolling down, rain falling, and many minor changes of this sort.

But there are many actions besides these. For instance, suppose before us a spring coiled up. When it unwinds it "exerts force," it transmits movement. In its first state it is like a stone at the top of a mountain. In its second state it is like a stone which has fallen to the bottom of the mountain. It had a power of movement and of communicating movement, now it has lost that power.

Again, the powder in a gun when it explodes expands and imparts movement to the shot. When the gun has been fired off the powder enters a different state. Before, the chemical affinities of its constituents were in a state of tension, now that it is fired off, they have formed fresh combinations. The power of transmitting movement has been lost by that which was the powder. It is like a portion of water at the top of a fall of water. If it remains at the top it has at any time the power of producing a shock, and of effecting, say, the movement of a water-wheel under it. But if it falls it has exerted and lost that power.

The difference of level associated with gravity is familiar to us. But we have no right, other than our own familiarity with it, to look on gravity as less in need of explanation than any other phenomenon of the external world. Newton did not suppose that there was any force inherent in matter which attracted other matter inversely as the square of the distance. He showed that a great many astronomical facts were capable of being explained and calculated on this hypothesis. He left the explanation of how it is that matter gravitates unsolved, and it remains unsolved to the present day.

But gravitation affords us a useful term—"Level."

Let us agree to call the following on a high level—a stone at the top of a precipice, a wound-up spring, oxygen and hydrogen mixed in the proportion to form water. Let us call the following at a low level—the stone at the foot of the precipice, a spring straightened so far as it tends to straighten, oxygen and hydrogen united in the form of water.

In passing from their first state to their last all these have manifested a power of movement and of communicating movement. They have now relatively to their former state lost that power.

Difference of level in this general sense is the most universal distinction of matter.

No motion takes place unless matter passes from a higher to a lower level.

The universal cause of motion is that which produces this difference of "level" in the general sense.

If there were no difference of level the state of things in nature would be as if one spring in order to unwind had to wind up an exactly equal spring of the same amount ; as if a stone falling from a height had to raise an exactly equal stone to the same height from which it fell. Under such conditions of things no motion would begin. In such a state of things all nature would be like the inhabitants of the valley when the king bore no pain, for no course would be preferable to any other course.

What is the cause of the " Difference of Level ?"

Whenever matter passes from a higher to a lower level some of the energy which is given out passes away in the form of heat. This passing away of some of the energy into the form of heat is an invariable accompaniment of the transition from a higher to a lower level. Is it the cause of the difference of level?

In the valley the king by bearing some pain made action worth while. Is there any indication in nature of the production of a lower level which makes the course of things run on.

It is certain that energy in every action passes off into the form of heat, and unless it is through the power of the finer particles of matter to absorb the energy, it is difficult to see how any action can take place.

As with the other lines of thought, this line also terminates with a possibility. Nothing has been proved, but a place has been provided.

In the first part of this paper a possible mode of action was exhibited in the imaginary relations of a world subject to certain laws of pleasure and pain.

In the second part it has been shown that something is wanting in our conception of the natural processes. There is room for a central idea. No scientific doctrines properly understood would clash with one properly located.

Can the mode of action exhibited in terms of sensation in the fictitious world be applied to the case of the world of force and matter?

Before passing on, however, it is worth while to examine a little more closely into what is meant by the expression so often used: " Passing off into the form of heat."

The modes in which energy passes off into the form of heat are in general those modes by which movement is brought to a standstill such as friction. And we are apt to think motion the primary fact, the cessation of motion a secondary and disagreeable fact. But both are equally existent phenomena, and the convenience to ourselves is not to mislead us as to their relative importance.

But what is this passing off of energy into the form of heat? The phrase is unsatisfactory, for we are told by science almost in the same breath that heat is the motion, the mechanical motion, of the particles of matter. So the statement resolves itself into this. Only when some of the motion passes off into the form of motion of the smaller particles of matter does motion take place in larger masses.

As a corollary it follows that at some date, however distant, all the motion of masses will have passed away into the form of motion of smaller masses.

It may be urged that when the larger masses move, the smaller particles also move. This is true; but motion in this sense is used to denote change of position amongst the smaller particles with regard to one another. The particles in a flying cannon-ball are relatively still with regard to one another as far as the motion of the cannon-ball as a whole is concerned.

We thus arrive at the following principle: The condition of the motion of masses taking place is that some of the motion passes off into the motion of the smaller particles.

But if the motion of the smaller particles is just the same as that of the larger portions, we are obviously not at the end. The very same principle just applied must be applied again.

These motions of the small particles of matter cannot take place unless some of their movement is transmitted and passed on, and transformed into the motion of still finer particles of matter.

But here obviously we are brought to the beginning of

an infinite series.   An infinite series passing from finer
matter to still finer matter, and so on endlessly.

The assumption by which we are led to this endless
series of transmissions must be clearly apprehended.
We take the law—that the motion of masses only takes
place when some of the motion passes off into the
motion of the finer particles of matter, and we assume
that it holds always.

In a lever there is a fixed point, the fulcrum, which
supports it, and the power raises the weight ; but the
weight may be fixed, and then the fulcrum can be lifted
by the power.   So we obtained this law from the consideration of material relations ; and now we suppose
this law to be the fixed point, and shift our notions
of material relations.

Thus we are landed on an endless series.   Before proceeding, however, to inquire what the significance of this
endless series may be, let us assume an end to it.   Let us
assume that we come at last to a final transmission.   Let
us assume that the energy is transmitted to the ultimate
particles of matter.

Or, if we have gone beyond matter, let us suppose an
ultimate medium which by its modifications builds up
matter, and which is the last and ultimate substance.

Let us suppose this ultimate medium absolutely to
receive some of the energy.   Let it absolutely receive
and absorb some of the energy, and thereby give rise to
the difference of level, to give the ultimate permission
which sets all things going.

What are the properties of this medium ?   We obtain
an indication of what they are when we examine the
properties of the finer kinds of matter.   Compared with
the motions of masses, motions which affect the smaller
particles of bodies are infinitely quick.   Light and electricity are actions affecting the smaller particles of

bodies, and by them distances are speedily traversed, which relatively to moving masses are very great.

Now in point of speed of transmission the properties of this ultimate medium must be infinitely beyond those of luminiferous ether.

To this ultimate medium all movements at any distance from each other must be almost equally present at every part. At whatever distance from one another two affections of this ultimate medium be supposed to take place, the effect of the one will travel to the place of action of the other instantaneously.

Such a medium is a kind of visible symbol of the universe, being one system in which all motions should be co-determined.

To make this clear, suppose a transformation of energy was produced in one part of space of an absorption of energy on the part of this ultimate medium, this transformation of energy would be produced by a medium in instantaneous contact with every other part of space, and the transformation of energy thus originated would harmonize with, and have reference to, the transformations of energy in every other part of space.

There are two infinities—the infinite of space extending out each way, the infinite of the smaller and smaller divisions of matter. The ultimate medium we have supposed partakes of both infinities. It is infinite in extent, and infinitely fine in its particles.

Now this medium by absorbing energy sets movements going. And that movements do not neutralize one another—*i.e.*, that movements in opposite directions do not mutually destroy one another—has this result, that a given amount of this absorption produces the greatest possible amount of motion. If motion came to a rest in any other way, more of this absorption by the ultimate medium would be needed. Hence, by a given amount

of absorption in the ultimate medium the greatest possible amount of motion is produced.   That is, the
absorption of motion into the ultimate medium is a
minimum, and the law of the conservation of energy is
the expression of this being a minimum.

But here again a further remark is called for. We start
by assuming energy to be an absolute existence.   But
why not assume this action on the part of the ultimate
medium to be the real action, and consider the phenomena of motion and energy as the mode of its action.

What this action of the ultimate medium may be
needs examination.   All that we can say at present is
that relatively to that which we call energy, the action
of this medium is that of being acted on.

―――――

CHAPTER VI.

In the preceding, however, it must be remembered that
this conception of an ultimate medium was merely a
supposition to enable us to see and roughly map out the
relations of the things we are investigating.   Where we
were really landed was in an infinite series—we were
brought logically to the conception of an infinite series
of media, one behind the other.

What does an infinite series indicate?

Let us turn to a region of thought where infinite series
are familiar objects, and we can learn about them.

In algebra infinite series are common.   Thus take

series $1 - \dfrac{x^2}{2} + \dfrac{x^4}{4}$ and so on for ever. This is the attempt

in algebra to represent a trigonometrical idea.   In trigo-
nometry it is expressed as cos. x.   But in algebra it
needs this infinite series.

In algebra infinite series occur when the object which it is wanted to represent in algebraical terms cannot be grasped by algebra. When there is no single term or set of them in algebra which will serve, the object is represented by means of an infinite series. Thus we may say that in any calculus, when the object to be treated of cannot be expressed in the terms of the calculus, it is represented by means of an infinite series.

Now, dealing with material considerations, going on in the calculus of matter, we have come to an infinite series. This indicates that we have gone as far as the material calculus will carry us. We have now to bring in an idea from a different quarter if we will simplify our expression.

It may well be that within our experience there is nothing which will serve. But let us suppose that that which in material terms we represent as an infinite series is a will—a will in contact with all existence, as shown by the properties it had when we conceived it as an ultimate medium. For, regarding it as an ultimate substance, we found that it would be affected by pulsations infinitely quicker than light and electricity; considered as a substance, it was such that distance to it tended to be annihilated. Hence as a will we must say of it that to it all that is is present—a will which by a fiat that to our notions is being acted on rather than acting, accepting pain rather than taking pleasure, sets the course of the world in motion, which holds all in one system, which creates all activities. For although we apprehend this will relatively to the appearances which we suppose we know, mechanical energy and feeling, still we see that both are caused by it, and that the sum of both is nothing, save for that which this will is in them.

Is there any other way of apprehending this will than through the external world?

We have two apprehensions of nature—one of external things, the other of our own wills.

Does this will not exist in those who are true personalities, and not mere pleasure-led creatures?—have they not some of this power, the power of accepting, suffering, of determining absolutely what shall be?—a creative power which, given to each who possesses it, makes him a true personality, distinct, and not to be merged in any other—a power which determines the chain of mechanical actions, of material sequences—which creates it in the very same way in which it seems to be coming to an end—by that which, represented in material terms, is the absorption of energy into an ultimate medium ; which, represented in terms of sensation, is suffering ; but which in itself is absolute being, though only to be known by us as a negation of negations.

CHAPTER VII.

IN conclusion let us remark that we have supposed two different worlds—one of sensation in the first part, one of motion in the second part. And these have been treated as distinct from one another. And especially in the first part, by this avoidance of questions of movement, an appearance of artificiality was produced, and occasionally inconsistencies, for sometimes sensations were treated as independent of actions, sometimes as connected with them. But it remains to be decided if these inconsistencies are in themselves permanent, or whether, when we remove the artificial separation, and let the world of sensation and the world of motion coalesce, the inconsistencies will not disappear, thereby showing that their origin was merely in the treatment, not in the fact ; that they came from the particular plan adopted of writing about the subject and are not inherent in the arguments themselves.

The king in the first part was supposed to have all the material problems of existence solved. There was a complete mechanism of nature. He took up the problem of the sentient life. But this problem can only artificially be separated from that of the material world. The gap between our sensations and matter can never be bridged, because they are really identical.

Let us then allow this separation to fall aside. Let us suppose the king to have all the reins of power in his own hands. Let us moreover suppose that he imparts his rays to the inhabitants so that they have each a portion of his power. And let us suppose that the inhabitants have arrived at a state of knowledge about their external world corresponding to that which we have about the world which we know.

Let us listen to a conversation between two of them.

A. The energy of the whole state of things is running down.

B. How do you prove that ?

A. Whenever any motion of masses takes place a certain portion of the energy passes irrecoverably into the form of heat, and it is not possible to make so large a movement with those same masses as before, do all that is possible to obtain the energy back again from the heat into which it has passed.

B. Well, what about the heat ? Energy in the form of the motions of the masses passes off into the energy of heat. But what is heat ?

A. It is the motion of the finer particles of matter.

B. Well, I would put forward this proposal. We have by observation got hold of a certain principle that where any movement takes place some of the energy goes in working on the finer particles of matter. Let us now take this principle as a universal one of motion, and apply it to the motions of the finer particles of matter themselves, which are simply movements of the same kind as the movements of the larger ones. This principle would show that these movements are only possible inasmuch as they hand over a portion of their energy to work on still finer matter.

A. Then you would have to go on to still finer matter.

B. Yes, and so on and on ; but to fix our thoughts, suppose there is an ultimate fine matter which is the last worked on. Now I say that we may either suppose that this is being gradually worked on and all the energy is dissipating, or else we may put it in this way. When we regard so much energy we are apt to think that it is the cause of the next manifestation in which it shows itself. But this is really an assumption. Energy is a purely formal conception, and all that we do is to trace in the actions that go on a certain formal correspondence, which we express by saying that the energy is constant.

A. But I feel my own energy.

B. Allow me to put your feeling to one side. If we take then the conservation of energy to be merely a formal principle, may we not look for the cause of the movements in the invariable accompaniment of them, namely, in the fact that a certain portion of the energy is expended irrevocably on the finer portions of matter. If now we take this ultimate medium which suffers the expenditure of energy on it, may we not look on it as the cause, and the setter in action of all the movements that there are. By its submitting to be acted on in the way in which it does submit, it determines all the actions that go on. For what is all else than a great vibration, a swinging to and fro. When we count it as energy, we by reckoning it in a particular manner make it seem to

be indestructible, but that the energy should be indestructible would be a consequence from the supposition
which we could very well make, that to produce a given
series of effects the submitting to be worked on of this
ultimate medium must be a minimum. If it were a
minimum no movements could neutralize one another
when once set going, for if they did there would be a
waste of the submission of this ultimate medium.

A. But what do you suppose this ultimate medium
would be?

B. That I cannot tell, but we seem to have indications.
For the more fine the matter which we investigate, the
more its actions seem to annihilate distance: light and
electricity produce their effects with far greater rapidity
than do the movements of masses. We might suppose
that to this ultimate matter all parts were present in
their effects, so that anything emanating from the ultimate matter would have the appearance of a system
comprehending everything.

A. But you have not got any evidence of an ultimate
matter.

B. No, all that we can think of is an endless series of
finer and and finer matter. But is that not an indication
rather, not that the direction of our thoughts is false, but
that there are other characteristics of this ultimate, so
that when looked at under the form of matter it can
only be expressed as an infinite series.

Let us omit the considerations brought forward in the
preceding conversation and examine more closely the
philosophy of the inhabitants of the valley in so far as it
corresponds with ours.

They laid great stress on a notion of *vis viva*, or what
we should term energy, but said it was gradually passing
away from the form of movements of large bodies to
that of movements of small bodies. So that in the

course of time the whole valley would consist of nothing but an evenly extended mass of matter moving only in its small particles—and this motion of the small particles they called heat.   Now they had very clearly arrived at the conviction that with every mechanical motion there was a certain transference of *vis viva* to the smaller particles of matter, so that it did not appear again as mechanical motion.   But they did not accept this as a principle to work by.   They did not consider that the motions of the smaller particles of matter were just the same as those of the larger masses.   They did not see that if a condition held universally for the movements of the visible world, it must also hold for the smaller motions which they experienced as heat.   So the conclusion which they should logically have come to that there was a transference of *vis viva* on and on was not held.   But the step was a very little one for them to take from regarding an invariable condition as always there to regarding it as a cause.   For the causes they assigned were all purely formal relations, and only got to assume an appearance of effective causes by familiarity with them, and a throwing over them of that feeling of effectiveness which they derived from the contact which they had with the king.

They might have reasoned.   This universal condition of anything happening must be the cause.   Energy goes from a higher to a lower level.   That which causes the difference of level is the cause, and the cause of the difference of level must be that which invariably accompanies such a transference of energy from a higher to a lower level.   Now this invariable condition is the passing of a portion of the energy into the form of motions of the finer parts of matter.   Hence there is an apparently endless series.   But to realize the matter, suppose an ultimate medium, suppose there is a kind of matter of infinite fineness distributed everywhere which let itself be worked on, and so determines differences and wakens the sleeping world. What are the qualities of this fine matter? We see them in the properties of the finer kind of matter which we know, such as light, electricity. The property of the finer kind of matter is in general that it tends to bring distant places together, so that a change in one part is rapidly communicated to every other part. If they followed this indication they would have supposed that the ultimate fine matter was of such a nature as to make all parts of the valley as one, so that there was no distance, and any determination of a difference of level on the part of this ultimate matter would have reference to all the conditions everywhere. It would be in immediate contact with every part, so that anything springing therefrom would present the appearance of a system having regard to the whole. Now if they had imagined such an ultimate medium doing that which to them would seem bearing rather than exerting force, suffering rather than acting, they would not have been far from a true conception of the king who directed them all. For he himself by reason of his very omnipresence could not be seen by them. There was nothing for them to distinguish him by. But they could have discovered somewhat of the means by which he acted on them, which can only be described from the appearances they present to the creatures whom the king calls into life.

But of truth they would have had another and perhaps a truer apprehension of the king in a different way. For when he acted on them so that they took one course rather than another, it was his action in themselves that they felt. If they were mere pleasure-led creatures then they were shaped outwardly, but if in their inner souls he acted and through them suffered, then they were true personalities conscious of being true selves, the oneness of all of them lying in the king, but each spontaneous in himself and absolute will, not to be merged in any other.

Thus they had two modes of access to the king, one through their own selves where he had made them exist, one through the outer world. And in the outer world it was but a direction in which they could look. They could never behold the personality of the king, but only an infinite series of different kinds of matter, one supporting the other as it were and underlying it, but doing more also than this, for in proportion as they considered the kinds of matter that lay deeper they found that distant became near, absent, present, that time gave no longer such distinctions, but from the phenomenal side they seemed by a gradual diminution of the limitations of experience to arrive at an external presentation of that absolute which exists in the fulness of things, which they knew more immediately in themselves when they truly were.

THE END.

INTRODUCTION.

I N the next two or three of these papers certain questions connected with the subject of a space higher than our own will be treated. It is well, therefore, first to recede and to form definite conceptions about a world of plane space, about a world in which the beings can only move in two independent directions. Then, proceeding thence to our own world, we may gain the means of passing on to a higher world. And I should have wished to be able to refer the reader altogether to that ingenious work, "Flatland." But on turning over its pages again, I find that the author has used his rare talent for a purpose foreign to the intent of our work. For evidently the physical conditions of life on the plane have not been his main object. He has used them as a setting wherein to place his satire and his lessons. But we wish, in the first place, to know the physical facts.

With this aim it is necessary to form a clear idea of what matter would be in a world of two dimensions, and the following illustration is a convenient one.

Place on the smooth surface of a table a half-crown piece, and suppose it to slide on the table perfectly freely. Imagine it to exercise an attractive force along the surface of the table in all directions round itself. By it and near it place a sixpence, and let the sixpence also slide freely on the table. It will, however, not be so free to move equally in all directions as the half-crown was, for it will be attracted by the half-crown. It will slip over the surface of the table under the influence of this supposed force, and will come into contact with the half-crown. Now if we suppose that both the half-crown and the sixpence are very thin, that they are both of them only the thickness of the ultimate particles of matter, then we shall have a representation of what material bodies will be in a plane world.

We must suppose that the particles cannot lift themselves or be lifted up from the plane so as to lie upon each other. Under no circumstances can they quit the surface of the plane.

Moreover, at no point must the particles adhere to the plane, nor must there be any friction impeding their movements over it. The only purpose which the support serves is to keep them on the same level surface and to convey influences from one particle to another. The gravity which we know, and which acts at right angles to the table on which the coins rest, will not have any effect on the particles in their motions on the plane, but will simply keep them to the plane. Any force of attraction which concerns their motions proceeds from one particle to another. Thus, conceive the half-crown to be a very large disk of matter, and the sixpence to be a sentient being. This being would feel a force of attraction towards the centre of the half-crown, and this force of attraction would keep him to the rim of the half-crown. If he weighed anything it would be by balancing it with his weight against the force which tended to pull it to the centre of the half-crown. He would not feel the gravity which keeps him against the surface of the table; he would not know that there was a hard, smooth surface on which he rested. He would always have been in contact with it, and so he could not tell what it would be like to be free from it. He would have no contrast whereby to apprehend its effect on him. Moreover, he would only know of movements in directions along the plane. He would not conceive that such a thing was possible as movement in another direction than to and fro, hither and thither on the plane. It is difficult to suppose that a being would be supported on one side by a plane, and not be in contact with anything on the other side, even atmosphere. Yet if we suppose a being of real matter free to move on the plane, this is what must be conceived. If the sixpence is conceived as such a being, it must receive its impressions through its rim. The rim represents its skin.

And if it be supposed to be surrounded by air for its respiration, this air must not be able, any more than the particles of solid matter, to rise away from the plane. The plane being must be conceived to have a different air to that which we know. The particles of its air, however free to move amongst themselves, must not have the power of moving away from the surface of the plane, as if so they would be able to pass to the interior of the body without passing through the skin. Any passage leading to the interior of the body would have to terminate in an opening in the rim, otherwise it would be completely shut up from the exterior.

Now it is obvious that if the table is struck so that it quivers, this movement will be communicated to the coins lying on it. Either the coins as a whole will move, or their particles will be disturbed.

Again, if we suppose there to be some particles loosely cohering together, lying on a smooth sheet of iron, it is evident that the quivering and jostling of the iron, if it is struck, would have an effect on the particles, and may cause the breaking up of the thin masses in which the particles cohere. Thus, if the material of which the

sheet is composed be very dense and rigid, compared
to the substances lying on it, they may undergo many
alterations, being broken up and coming together again,
while the supporting matter which bears them all up
simply moves and vibrates.

It is evident that just as the particles are affected by

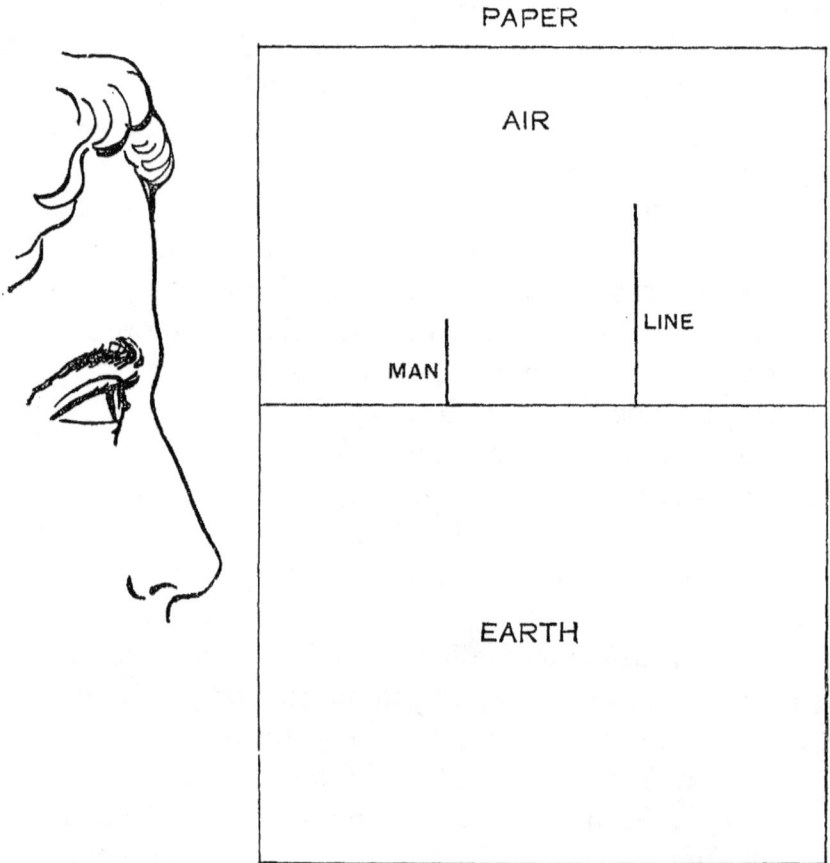

Diagram I.

the vibration and shaking of the sheet of metal on which
we suppose them, so they might in turn possibly affect the
sheet of metal and cause vibrations and shakings in it.
These shakings and vibrations would go forth from a
particle which excited them in every direction along
the sheet. They would not pass out into the air, except

secondarily and in a very minute degree. The shake would be transmitted in the sheet. And the effect on neighbouring particles would be great, on more distant particles it would be less, and on those at a great distance barely perceptible.

The following is a good plan for obtaining in a definite way the feeling of what existence in a plane would be like ; it enables us to realize the conditions in such a way as to lay the basis for subsequent thought.

Let the reader take a sheet of note-paper and hold it before himself edgewise, so that he sees it with one eye as a single line. And let him hold it so that this line runs downwards from his eyebrows to his mouth, as shown in Diagram I. Now on this sheet of paper, on one side of it, let a straight line be drawn running across it, away from the observer. Suppose all below this line to be a thin layer of particles which, keeping compactly together, form a solid sheet of particles, every one of which touches the paper. This would be the solid earth to a being in the plane world.

Let the surface of the paper above this be covered by a layer of particles which move freely amongst each other, but which do not rise from the surface of the paper. These particles form the air of such a world.

On the surface of the earth draw a line standing upright. Let this line represent a man. Another line will represent a wall which the man could not pass except by getting over it.

It will be found that the objects on the paper are felt to be subject to the action of gravity. The question will occur, Why will not this thin layer of particles slip off the paper ?

Now, the sense of gravity must not be got rid of, but it must be connected with the matter in the sheet of paper.

Suppose, then, that the sheet were to grow bigger and bigger till it filled out reaching through the whole world and cutting the globe in two. Then let all the earth be removed except a thin layer on one side of this enlarged sheet of paper. This thin layer will be the only portion of matter left. And such a thin layer will represent a plane world. The force of gravity must be conceived as remaining, but as coming from a large and thin disk.

Now to keep this thin layer on the paper it would be necessary to have some force acting sideways, so as to keep the particles to the paper.

And the paper itself may be conceived to exercise such a force : it is many particles thick, while the thin layer of matter is only one particle thick, and thus it will keep the layer of matter, which covers one side of it, in its place by virtue of its own attraction.

We suppose that the paper exerts an attractive force which keeps the thin layer of matter to it. This attractive force is not felt by the sentient beings on the paper, nor does it influence the movements of the particles of matter amongst themselves. We also suppose another attractive force proceeding from particle to particle of the matter on the plane. This would be felt by the beings and produce movements of matter.

Thus the conception of a plane world necessarily involves that of something on which it is.

A Plane World.

WHERE the sun's rays grazing the earth in January pass off and merge into darkness lies a strange world.

'Tis a vast bubble blown in a substance something like glass, but harder far and untransparent.

And just as a bubble blown by us consists of a distended film, so this bubble, vast beyond comparison, consists of a film distended and coherent.

On its surface in the course of ages has fallen a thin layer of space dust, and so smooth is this surface that the dust slips over it to and fro and forms densities and clusters as its own attractions and movements determine.

The dust is kept on the polished surface by the attraction of the vast film; but, except for that, it moves on it freely in every direction.

And here and there are condensations wherein have fallen together numbers of these floating masses, and where the dust condensing for ages has formed vast disks.

And these disks are glowing hot—yet no light comes from them into our universe.

For this world lies beyond the æther—far beyond. And however hot or glowing the masses are, if there is no medium to transmit the vibrations of heat the influence cannot travel.

Thus the only directions in which the heat can travel

are on the film. From each of these glowing disks the luminous influence streams forth carried by the vibrations of the film which supports everything. For the heat and intense agitation of these glowing disks shakes and disturbs the bubble, and just as a thin soap bubble quivers and shakes, so this film quivers and shakes. And so elastic is it, and so rigid, that it carries the light and heat to all surrounding regions. Yet so vast is the bubble, so tremendous in its dimensions, that the agitation from these glowing disks travels almost in straight lines, till, spreading out on every side, it merges into darkness— like the ripples in the centre of a vast calm lake gradually become indistinguishable.

And round these central orbs of fire—for orbs of fire they are, though they only transmit their fire along the film of the bubble—round these orbs pass in due order and succession other disks, which, cold or warm, have not that energy of light and heat which the central orbs possess.

These disks, though large, are so immeasurably small compared with the vast surface of the all-supporting bubble, that their movements seem to lie on a plane flat surface; the curving of the film on which they rest is so slight compared to their magnitude, that they sail round and round their central fires as on a perfect level surface.

And one of these orbs is fitted by nature to be the habitation and home of living beings. For it is neither so hot as it was for long ages after it had condensed from the film of dust wherefrom all orbs are made, nor has it so cooled down as to render life unsupportable.

And, moreover, it is full of vast crevices and channels, for in many places the interior in cooling after the rim had set from its molten condition has left long caverns and passages, not only in one layer, but in many.

And on the rim and in these passages and caverns
live the inhabitants of whom I speak.

They do not rise from the surface of the film, but as
all matter lies on the smooth surface but one particle
deep, so their bodies formed out of matter lie, as we
should say, on this smooth surface.

Yet of this they know nothing. They say that they
stand and walk.

For this orb has an attractive force.

By that very same impulse of coming together whereby
it gathered its particles out of the dust on the bubble,
by that very same force it draws towards its centre all

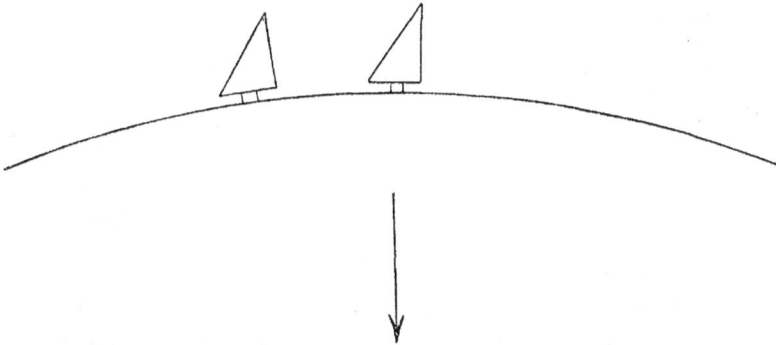

Diagram II:—" Two beings walking round."

that is near it or on it. Thus " up " is to these inhabitants a movement from the centre of the disk on the rim
of which they live and away from it. " Down " is a movement from the rim towards the centre. The thin layer
which forms the mass of the disk is their solid matter.
They are not able even in thought to rise away from the
surface of the bubble, and look from space upon their
mode of existence. They ever pass to and fro upon a
line, upon a rim ; and no two can walk except after one
another. If you look at the rude picture you will see
that the two beings represented by two triangles cannot
pass one another if they are unable to lift themselves up
from lying on the surface of the paper. The surface of

the paper represents the surface of the bubble, and sliding freely on it, but unable to lift up from it, are tenuous shapes that are the inhabitants, and that thin layer of particles that is to them solid matter.

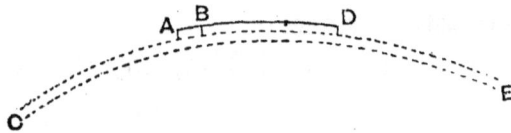

Diagram III.—A section of the film of the bubble showing a disk BD lying on it, and a creature AB on the rim of the disk. CE is a section of the film, BD is a section of the disk, AB is a section of the creature. The thickness is enormously magnified and also the height AB of the creature compared with the diameter BD of the disk. The attraction which AB feels keeps him to BD ; both AB and BD, the being and the disk, slide freely on the film CE without knowing of its existence.

Now were it not for the fact that the orb is reft into these chasms and passages, the only movement that these beings would have would be of passing round and round on the rim of their world.

Many words that we have, to them could bear no meaning. Thus "right and left" is to them unknown. For consider their faces bent in one direction along the rim. In following this direction, they go forward, in retracting from it they go backward. If they go away from the centre they go up, towards the centre is down. And by no means can they turn, raising themselves from the surface whereon they are. They do not even know that they have two sides; their movements, thoughts, and imaginations are all confined to that surface on which they are. This they call their space, their universe; nor does aught that lies beyond it, towards the interior of the bubble or away from it, directed outwards, come into their thoughts, even as an imaginary possibility of existence.

Life is extremely limited on such a world. To take a single instance, in order for two beings to pass each other, a complicated arrangement is necessary, shown in Diagram IV.

At intervals along the rim recesses and chambers are

constructed. Near the openings of these chambers lay movable plates or rods. When two beings wish to pass, one of them descends into the recess; the other one pushes the rod so as to form a bridge over the opening, walks across it, and then removes the plate so

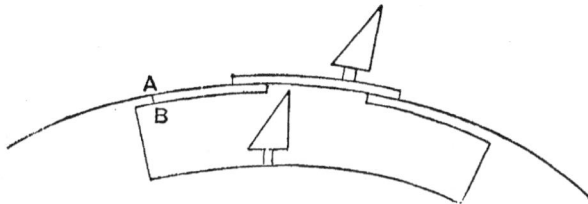

Diagram IV.—Two beings passing.

that the one who has descended can get up and go on his way.

If by any chance, while a being is in the recess, the plate or rod which acts as a bridge gets fixed, he is in a dangerous predicament. For suppose a being confined as shown. If he, suffering from want of air, cuts through the roof at AB, the whole part to the right of AB comes tumbling down. For its only support is severed when AB is cut through. It is impossible to make a hole which is not the whole width of matter as it lies on the surface. And with regard to this all constructions have to be made. There cannot be two openings in a wall of a house, unless when the one is open the other is so fashioned as closing to act as a rigid support to the wall,

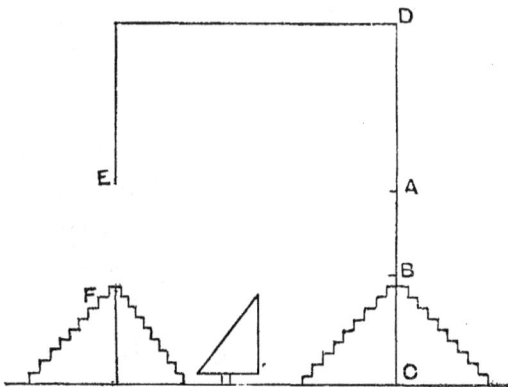

Diagram V.—A house.

which now depends for its upholding entirely on it.

Thus, in the diagram, the house is held up entirely by the side opposite to the doorway EF, which is now open. The roof is supported by the side CD. If an opening

AB be made in the wall CD before the doorway EF is closed, the roof will fall in. So, in order to pass through the house, EF must be firmly shut up before AB is opened. The houses are always built in the interior passages so as to leave the rim of the disk free for locomotion.

And there are many things to be said of the inhabitants on this disk with respect to their social and political life. It is hardly necessary for me to put down much about it here, for any one by using the method of the historian Buckle, and deducing the character of a people from their geographical influences and physical surroundings, could declare what the main features of their life and history must be.

But one or two remarks may be made here. First of all they are characterized by what I may venture to call a crude kind of polarity.

In dwellers in our world this polarity, which shows itself amongst other ways in the distinction of sexes, is tempered and modified.

In every man there is something of a woman, and in every woman there are some of the best qualities of a man.

But in the world of which we speak there is no physical possibility for such interfusion. In a linear existence there would be no consciousness of polarity. It makes its appearance first in the plane, and in a hard and unmitigated form.

It is impossible to do otherwise than caricature these beings when we write of them thus in brief. So let us accept the matter frankly, and, without scruple, look at them in the broadest possible manner.

If the reader will cut out the triangles in the corners of the two next pages he will obtain four plane beings, two of which are men, two of which are women. The

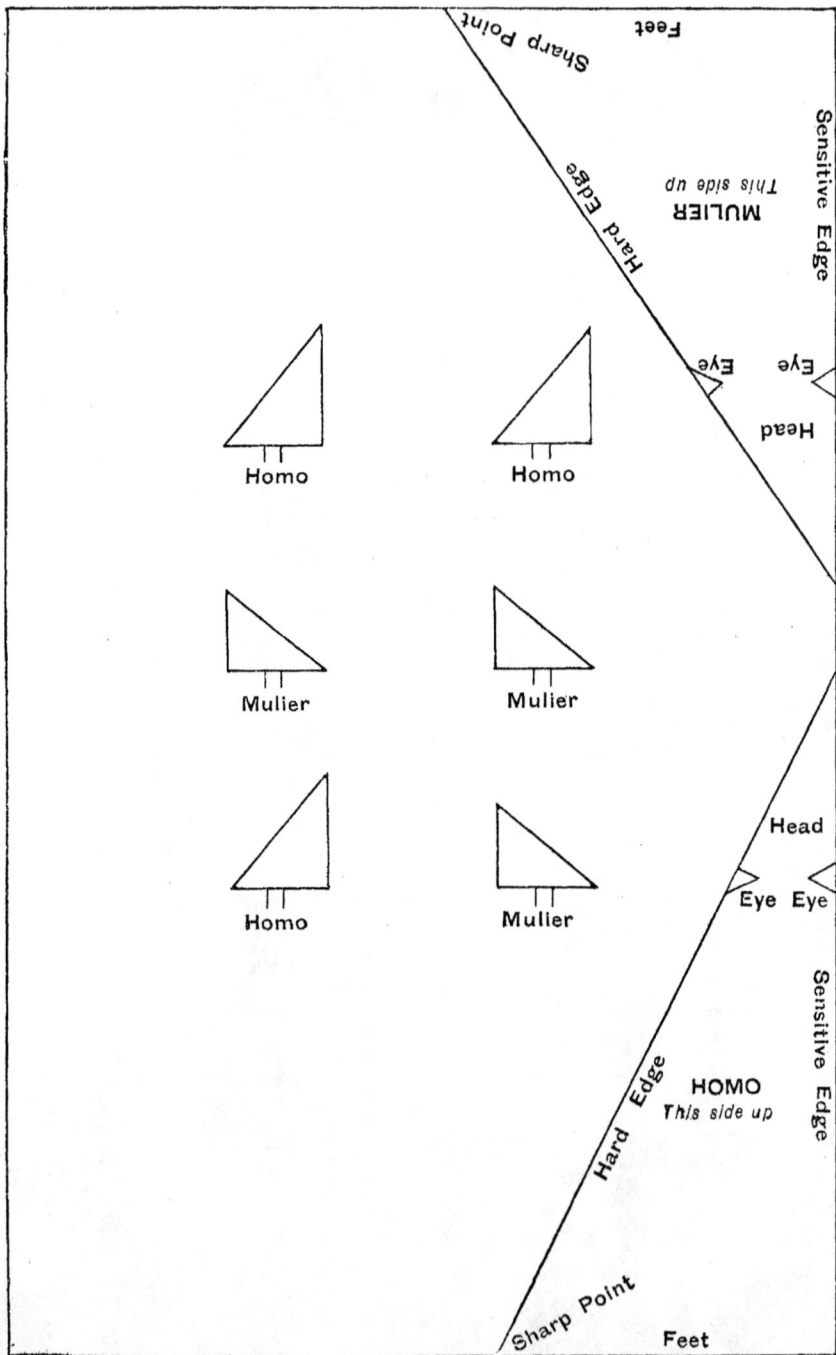

Diagram VI.

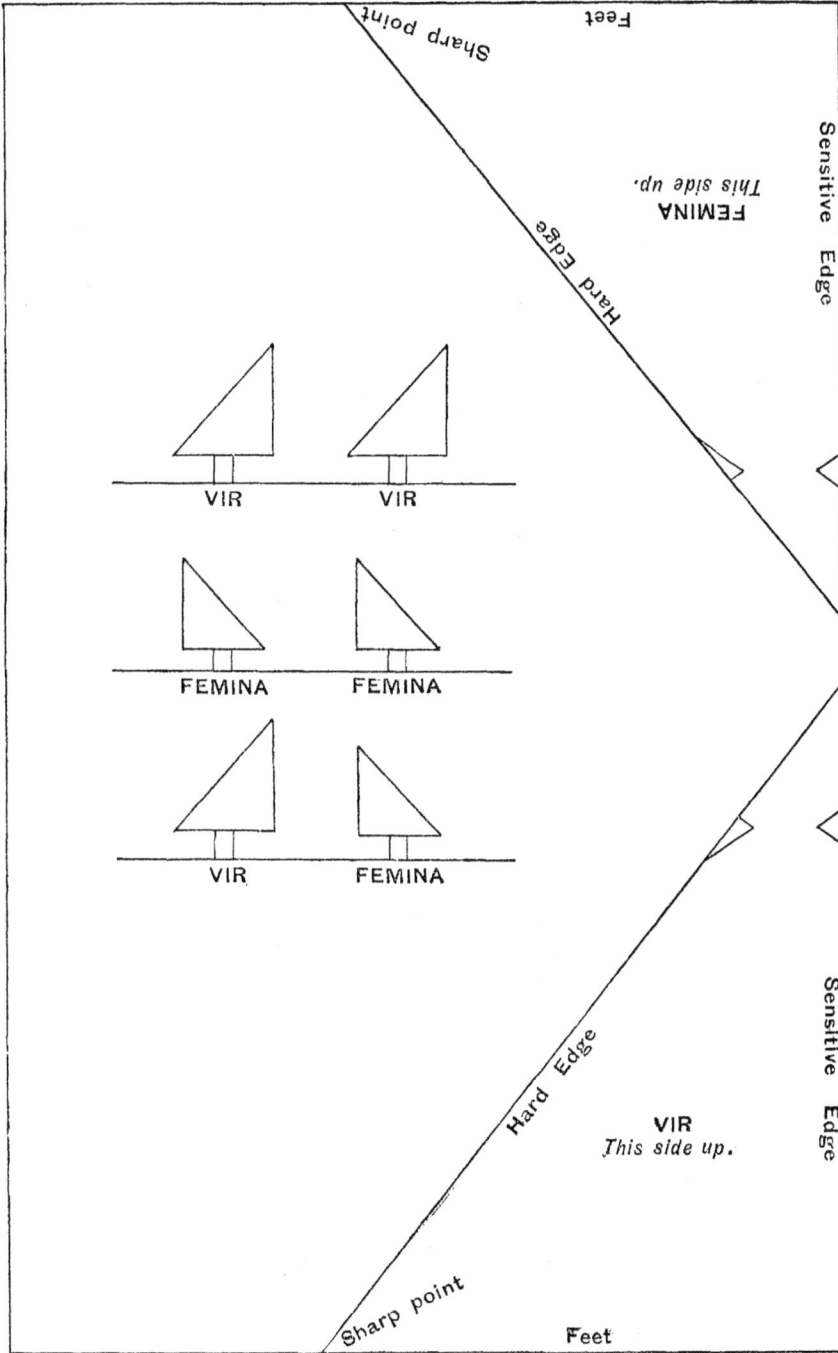

Diagram VII.

lines down which the cutting is to be made are marked
with a black line. Now having cut out two men, whom
we will call Homo and Vir, draw a line on a piece of
paper to represent the rim of the world on which they
stand, and, remembering that they cannot slide over each
other, move them about. It must be remembered that
the figures cannot leave the plane on which they are put.
They must not be turned over. The only way in which
they can pass each other is by one climbing over the
other's head. They can go forwards or backwards.
Much can be noticed from an inspection of these figures.
Of course it is only symbolical in the rudest way, but in
their whole life the facts which can be noticed in these
simple figures are built up and organized into complicated arrangements.

It is evident that the sharp point of one man is always
running into another man's sensitive or soft edge. Each
man is in continual apprehension of every other man :
not only does each fear each, but their sensitive edges—
those on which they are receptive of all except the
roughest impressions—are turned away from each other.

On the sensitive edge is the face and all the means of
expression of feeling. The other edge is covered with
a horny thickening of the skin, which at the sharp point
becomes very dense and as hard as iron. It will be
evident, on moving the figures about that no two men
could naturally come face to face with each other.

In this land no such thing as friendship or familiar
intercourse between man and man is possible. The
very name of it is ridiculous to them. For the only
way in which one man can turn his sensitive edge to
another man is if one of them will consent to stand on
his head. Fathers hold their male children in this way
when little, but the first symptom of manhood is connected with a resentment against this treatment.

If now two women, Mulier and Femina, be looked at, the same relation will be seen to hold good between them. By their nature they are predisposed, by accident, to injure one another, and their impressionable sides are, by the very conditions of their being, turned away from each other.

If now, however, Homo and Mulier be placed together, a very different relationship manifests itself. They cannot injure one another, and each is framed for the most delightful converse with the other. Nothing can be more secure from the outside world than a pair of approximately the same height ; each protects the sensitive edge of each, and their armoured edges and means of offence are turned against all comers, either in one direction or the other. But, if the pair, through a mutual misunderstanding, happen to be disadjusted, and, their feet on the rim, turn their sharpnesses against one another, they are absolutely exposed to the harms and arrows of the world.

Still, even in this case they cannot wound one another —a happy immunity.

In the annals of this race which I have by me I find a curious history, which, unintelligible for ages to them, admits by us of a simple explanation.

It is said that two beings, the most ideally perfect Vir and Mulier, were once living in a state of most perfect happiness, when, owing to certain abstruse studies of the Mulier, she was suddenly, in all outward respects, turned irremediably into a man. Vir recognized her as the same true Mulier. But she occupied the same position with regard to him which any other man would. It was only by standing on his head that he could, with his sensitive edge, approach her sensitive edge. She refused to explain how it was, or impart her secret to any one, but she had, she said, undergone a great peril.

She manifested a strange knowledge of the internal anatomy of the race, and most of their medical knowledge dates from her. But no persuasion would induce her to reveal her secret ; all the privacy of existence would be gone, she said, if she revealed it. She was supposed to have acquired some magical knowledge.

This possession, however, did not make either of them happy, and one day, with fear, she said that she would either die or be restored to the outward semblance of her sex.

She disappeared—absolutely ; although she was surrounded by her friends, she absolutely vanished. And had it not been that some days afterwards, cutting through the solid rock for the purposes of some excavations, they accidentally came on a chasm, they would never have found her alive again. For she was found in a cavity in the living rock, warm and beautiful—her old self again.

Her secret died with her.

From our point of view it is easy to see what had happened. If the figure Mulier be taken up and turned over it will be easy to see that, though still a woman, her configuration has become that of a man. To all intents and purposes she is a man. She is rendered incapable of that attitude which is the natural one between the men and women in this land, and the happy relationship between her and Vir is necessarily and entirely broken off. Move about as you will, keeping her figure turned thus on the plane, you will not be able to make her a fitting helpmate for her unfortunate Vir. She must have discovered the secret of raising herself off the surface, and by some accident been turned over. Perhaps she had used this new position to study anatomy—for to an observer thus situated the interior of every body would lie perfectly open—and in prosecuting her studies had overbalanced herself.

I have only mentioned this anecdote, however, for the sake of a curious observation which was made at the time. It was found that when she was in this transformed condition she was absolutely without atmosphere. To explain: ordinarily, apart from anything she said or did, there was a kind of influence proceeding from her which made her presence agreeable to Vir. When she was turned over she lost this. Now the explanation of this is obvious. To these people light is the agitation of the surface of the bubble; transparent objects are those which do not hinder this agitation in its course. But most bodies and the physical frame of the inhabitants amongst them were not transparent, but stopped and reflected these agitations of the film, thus sending off from their outer edge those vibrations which excited sight in their fellows. But besides these vibrations of light there were finer ones still which were not damped or deflected by the outer edge of the body, but came through the greater part of their frame as if it was transparent. In the interior, however, of their organizations there were certain regions which did arrest these subtler vibrations, and which had (as the eye of light) the power of appreciating them. In connection with these regions there were certain structures, extremely minute, which had the converse power of agitating the film, and so sending forth through the periphery of the body these same minute vibrations. These organs were not of any use, but they formed a sort of means of sympathetic communication between the inhabitants, acting in no very defined way, but certainly producing a sensation of a vague kind. Now when Mulier was turned in the way described, the relation of her frame to the film of the bubble was disarranged, and it was no wonder that this "atmosphere" disappeared.

In many respects the inhabitants of this world are far more advanced than we are, having a simpler problem— how to deal with matter in one plane—they have advanced more nearly to a complete knowledge of its properties. Yet great as their knowledge is, their performance is small. If you but reflect on one single fact, you will see how limited all their efforts must be. *They cannot fix the centre of a wheel, so that it rotates round an axis.* For consider a wheel —a small disk lying on their plane. The centre on every part of it touches the surface of the bubble on which all things slide freely. To fix this point they would have to drive down into the film—a thing which they cannot do, and which they are far from even imagining.

If they make an opening in the disk they can arrive at the centre of it. But then the rod of matter which they put in will prevent the disk from revolving.

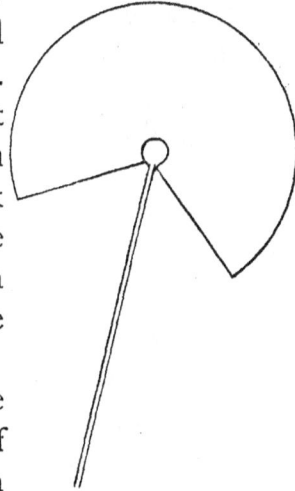

Diagram VIII.—The nearest approach to a wheel.

The nearest approach to a wheel with a fixed centre which they can attain is shown in Diagram VIII., a portion of a circular disk which oscillates about the smooth end of a rod built into the substance of the cut-away disk.

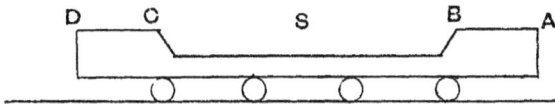

Diagram IX.—A cart.

Their carts are shown in the accompanying figure. They are simply rods placed on rollers: as the rod is pulled along, the rollers turn, and the rod slides along— just as a boat does on the rollers whereby sailors help themselves in hauling it up the beach. As soon as these rollers roll from under the rod, as it goes on in its forward motion, they have to be secured, and then lifted over the cart and put down in front. Thus there are to each cart a set of little disks or rollers, and, as the cart goes on, these rollers have to be lifted over the cart from the back to the front.

There is no means by which this can be made a continuous action. Each roller has to be waited for, lifted separately, and carried over separately. And to put it down in front, the rope by which the cart is dragged along has to be unfastened and fastened up again.

Looking at the Diagram IX. it will be seen that there is a hollow in the body of the cart. On the part A B the driver sits. In the hollow from B to C is put the load. The load cannot then slip out over the ends of the cart. There is nothing in the cart to prevent it from falling out sideways.

But the contents, as the whole body of the cart, are kept to the smooth surface of the bubble, and are thus supported by it on the side remote from the reader's eye, and also are kept from rising away from this surface by the force of attraction exerted by the film.

Thus the surface of the bubble and its attractive force supply the other two sides of the cart.

But of these two sides, the beings are ignorant, and it seems to them perfectly natural that loads of any kind, even of fluids, should be kept securely in a cart with two ends.

The method by which the rope is fastened to the cart is this : C is the body of the cart ; R is the rope ending in a wooden step B ; A is

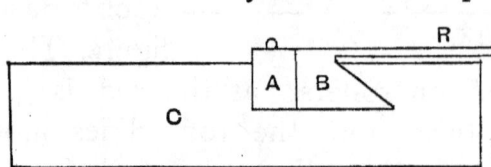

Diagram X.—How a rope is fixed to a cart.

an oblong piece of wood. When the rope has to be taken out, A is lifted out by its handle, B is slipped back and taken out of the recess in C, and then the rope is free from the cart. And in a similar way it is secured again.

One very ordinary way of driving machinery with us is by shafting. A long rod is driven round and carries wheels at different places along its length. Now with these inhabitants it was impossible to do this, because the twisting motion round a rod could not be imparted without going out of the thin layer in which they were. Their methods of transmitting motion were by long rods, by a succession of short rods, by pendulums attached to one another, or finally by wheels which drove one another, but which were held by smooth sockets fitting round the rim far enough to steady them, but not so far as to hinder them from touching each other.

As to their science, the best plan is to give a short account of its rise.

They discovered that they were on a disk rotating round an inner centre, and also proceeding in a path round the source of light and heat.

They found that they were held in their path by a force of attraction. But this attractive force was not with them as it is with us. With us, since the effect which any particle has on the surrounding particles spreads out in our space if the distance is doubled from a centre of attraction, the force it exercises becomes one-quarter of what it was when at the less distance.

With them, however, when the distance doubled, the force of attraction became only one-half of what it was at the less distance. For the light, or attraction, or force of any kind emanating from a particle, only spreads along the film, and does not pass out into the space above or beneath. If they had been on a thick globe instead of a bubble, the laws of attraction would have been the same as with us. But the bubble on which they were was thin compared with the paths along which the radiant forces spread forth. And thus every force being kept to one plane diminished as the distance from the centre of its action.[1]

Now it was a great problem with them how the light came from the central orb. Their atmosphere, they knew, extended but a small distance above the surface of their disk. And it was quite incapable, moreover, of conveying vibrations such as those of light and heat.

By studying the nature of light they became convinced that to transmit it there must be a medium of extreme rigidity between them and the great source of light.

It is easy enough to see that what they thought was a medium between them and their sun was in reality the rigid surface on which they rested. This elastic film vibrated in a direction transverse to the layer they called matter, and carried the particles of matter with it. But they, having no idea but that the surface on which they were was the whole of space, thought that space must be filled with a rigid medium. They found that the vibrations of the medium were at right angles to the direction in which a ray was propagated. But they did not conceive of a motion at right angles to their plane ; they thought it must be in their plane.

It was a puzzle to them how their disk glided with so little friction through this medium. They concluded it was infinitely rare. They were still more puzzled when they had reason to believe it was an opaque substance; and yet that it could be anything else than a medium which filled their space was inconceivable to them. They could never get rid of it from a vacuum, however perfect. Indeed we see that in producing a vacuum they merely cleaned the surface on which they were.

In one respect it might have been advantageous if

[1] *See* Appendix.

they had known, for, their law of attraction being what it was, their movement round their sun was not destined to go on for ever ; but they were gradually falling nearer and nearer. Now, if only they had made the attempt, they might by some means have got a hold on the surface on which they were, and, by means of a keel which tended to furrow it, have guided their world and themselves in their path round their sun. Indeed, it is possible to imagine them navigating themselves whither they would through their universe—that is, on the surface of their bubble.

It was also unfortunate in another respect that they did not realize the fact of the supporting surface, for the feeling which they came to have of being suspended in space, absolutely isolated, was a very unsettling one, and tended to cause in them a certain lack of the feeling of solidarity with the rest of the universe.

We have seen that their laws of mechanics were very different from ours. But they had after all an experience of our mechanical principles, though in a curious way. In all motions of any magnitude moving bodies were confined to the surface of the plane. But where the small particles were concerned there was more liberty of motion. The small particles were free in their movement ; although they could not go more than a very small distance away from the film on which they rested, still they were capable of motion perpendicular to it. Thus a long line of particles connected together could rotate as a whole, keeping straight like a twisting wire, and by means of many strings of particles thus connected, movements could be transmitted in a way which was totally unlike the mechanical movements to be seen in the case of large masses.

This motion of rotation round an axis lying in the plane was to them what electricity is to us. It was quite a mysterious force. But it was extremely useful in its applications. Having no idea of a rotation which in taking place went out of their surface, they could not conceive a reason for the results of such movements.

It can easily be seen how many kinds of forces they could have. There was the spinning motion of the small particles on the surface. This they were aware of—it produced many appearances, but it was not fitted for transmission across great distances, as each particle was apt to be hindered in its rotation by its neighbour. Sometimes, however, when conditions were favourable, many of these rotations were harmonious, and waves were produced in their matter resembling the waves in our ocean.

There were only two other kinds of motion. One was an up and down vibration of the film carrying matter with it; the other was the twisting of strings of particles which were rigidly connected together. The up and down motion of the film was to them light. Those kinds of matter which did not hinder this motion were said to be transparent; those kinds which, lying on the film, hindered the motion or threw it back were said to be opaque.

The twisting motion round an axis was to them what electricity is to us. And when this twisting motion in one direction or another was conveyed to the particles of small masses which were free to move, many curious effects were produced analogous to the movements of electrified bodies. There are obviously no other rotations or vibrations possible; hence in that world there is nothing corresponding to magnetism. Their light was simple, and could not be split into two kinds as our light can be—into two kinds of polarized light.

Was there no sign, then, by which the inhabitants of this world could gain a knowledge of their own limitation ? There was. There was both a sign and the interpretation of it lying before them. They knew that they could have two triangles precisely similar, and yet such as could not be turned the one into the other by any movement in the plane. How two things could be so alike, and yet differ in some mysterious way, was to them a puzzle. As an instance of such triangles may be taken those used in Diagram VI. to represent the man and the woman. They may be exactly equal, yet the beings in a plane world cannot turn them so that one would coincide with the other.

Yet had they but considered the case of a being lower in the scale of space existence than themselves, they would have seen the answer to their riddle. For consider a being confined altogether to a line

C'   B' A'   M   A B   C.   Let M be the being, and

let him observe the three points A B C, and let him form an idea of them and their positions with regard to each other, which he measures by the distance he has to travel to reach one after passing the other.

Let him also become aware of the three points A' B' C', forming a precisely similar set on the other side of him.

It may be objected that the being in the line could not conceive any point lying beyond A, but that his experience would be limited to the points A and A'. If A and A' are material particles this would be the case, but we may suppose them to be places in the line marked out by cold and heat, or some such means. Then a being could conceive a series of positions in his space such as A, B and C, A', B', C'.

If now he remembers each set, and thinks about them, he finds that they are alike in every respect. But he cannot make them coincide with each other. For if he pushes the set A B C along the line, when A B and A′ B′ are together C is just where it ought not to be. It is not on C′. And if he gets C on C′, then A B has gone far away.

He would neither be able to make them coincide nor to conceive their coincidence.

There would be no movement within the realm of his experience which would make them coincide.

Yet the dweller in a plane world could easily make these sets of points coincide, for he would bend the whole line round in his plane so that A coming on A′, B should come on B′, and C on C′. There would be no difficulty to him in doing this. And he does it in virtue of there being to him a movement possible which is not possible to the being in the line. He has a liberty of motion unknown to the linear being.

And now why should he not reason thus, " Something which to the linear being is inconceivable, to me is conceivable. Then may not things inconceivable to me be yet possible ? May it not be possible that two triangles which are like one another, but yet which cannot be thought by me as coinciding—may not these triangles be able to be made coincident "?

In this simple fact of his perpetual observation was really the proof of the whole matter if he had but looked at it, the sign manual of his limitation, the promise of his liberation from it in thought, the key to the explanation of the mysterious minute actions by which he was surrounded, and perchance a help to the comprehension of a higher life.

APPENDIX.

———◆◇◆———

IN our world a particle of matter which sends forth influence on the surrounding matter does not send its radiant energy off along a plane, but from the particle all the influence spreads out into space. And the most convenient instance in our world to consider is that of a luminous point from which rays spread out in every direction. Let M in Diagram XI. be such a point—a particle of

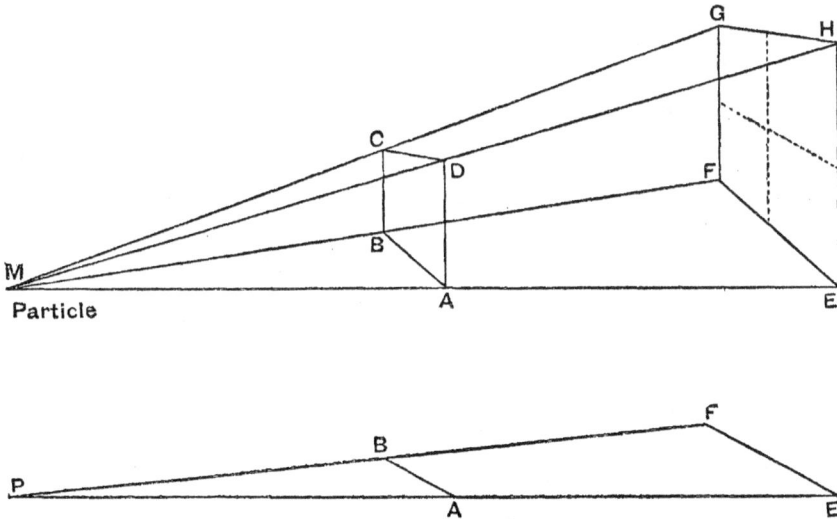

Diagram XI.—Particles in space and in a plane exerting force.

matter sending forth luminous rays in our three-dimensional space.

Instead of studying how these rays spread out in every way all around M, let us only consider those which, pass-

ing out from M, fall on the square A B C D. A B C D
casts a shadow, and this shadow extends, and is found to
be bigger the further off from M it is measured. Suppose,
at the distance from M, M E, we put a square in the path
of the shadow so as just to receive the shadow on it
exactly. Let E F G H be this square. As is shown
by the dotted lines, this square will be four times as
large as the square A B C D. So when the distance is
doubled, the shadow becomes four times as big.

Now those rays of light which fall on A B C D would,
if they were not interrupted by it, spread out so as to
exactly cover E F G H. Thus the same amount of
light which falls on the small square A B C D would, if
it were taken away, fall on the large square E F G H.

Now since the large square is four times the size of
the small square, and the same amount of rays fall on it
—for it only receives those which would fall on the small
square—there must be at any part of it an illumination
one-quarter as strong as there would be at any point on
the small square.

Thus the small square, if placed in its position, would
seem four times as bright as the large square.

Thus, when the distance from the origin of light is
doubled, the amount of light received by a surface of
given area becomes one-fourth of what it was at the less
distance.

This is what is meant by varying inversely as the
square of the distance. When the distance is doubled
the intensity of the light is not simply less, but is halved
and halved, and becomes one-quarter of its previous
intensity.

But in the case of a particle resting on a thin sheet of
metal, and shaking the metal—as, for instance, a metal
plate can be made to shake by a violin bow—then this
law would not hold.

Take the second figure. Let P be the particle, and let the influence proceeding from it fall on the rod A B lying on the plane, and let us suppose the rod to stop the vibrations from going beyond it, to receive them and to turn them back just as a body does the light. Then the "shadow" of A B would spread out away from P; and if another rod E F were put in at the distance P E, which is double of P A, then, to exactly fit the shadow, it would have to be double the length of A B; and the vibrations which fell on A B would exactly fall on E F. Now since E F is twice as long as A B, the vibrations which fall on any part of it will be one-half as intense as the vibrations which fall on a portion of matter of the same size lying where A B lies.

Thus in a plane the influence or force sent out by any particle would diminish as the distance. It would not "vary inversely as the square of the distance," but would "vary inversely as the distance."

A Picture of Our Universe.

CHAPTER I.

IT seems to me that the subject of higher space is becoming felt as serious, and fraught with much that is of the deepest interest, not only as a scientific problem, but in other ways also.

It seems also that when we commence to feel the seriousness of any subject we partly lose our faculty of dealing with it. The intellect seems to be overweighted somehow, and clogged. Perhaps the suppositions we make seem to us of too great importance, and we are not willing enough to let them go, fearing to lose the thing itself if we lose our hold of the means by which we have first apprehended it.

But whatever may be the cause, it does seem undoubtedly the fact that the mind works more clearly and more freely on subjects which are of slight importance.

And I propose, that without ignoring the real importance of the subject about which we are treating, we should cast aside any tension from our minds, and look at it in a light and easy manner.

With this object in view let us contemplate a certain story which bears on our problem.

It is said that once in a certain region of Ireland there took place a curious contest. For in Kilkenny

there were two cats so alike in size, vigour, determination, and prowess, that, fighting, they so clawed, scratched,
bit, and finally devoured each other, that nothing was
left of either of them save the tail.

Now, on reflecting on this story, it becomes obvious
that it originated when looking-glasses were first im-
ported into Ireland from Italy.   For when an Irishman
sees for the first time anything new, he always describes
it in an unexpected and yet genial and interesting
manner.   Moreover, we all know what contentious fellows they are, and how all their thoughts run on fighting.   And I think if we put this problem to ourselves,
how by bringing in fighting to describe a looking-glass,
we shall see that the story of the Kilkenny cats is the
only possible solution.   For consider evidently how it
arose.   Depositing his favourite shillaly in a corner, the
massively-built Irishman, to whom the possession was a
novelty, saw reflected in his looking-glass the image of
his favourite cat.   With a scrutinizing eye he compared
the two.   Point for point they were like.   " Begorra if I
know which of the two would win ! " he ejaculates.   The
combat becomes real to him, and the story of the Kilkenny cats is made.

Now, to our more sober mind, it is obvious that two
cats—two real material things—could not mutually annihilate each other to such an extent.   But it is perfectly
possible to make a model of the Kilkenny cats—to see
them fight, and to mark the issue.

And I propose to symbolize or represent the Kilkenny
cat by a twist.   Take a pencil, and round it twist a strip
of paper—a flat spill will do.   Now, having fastened the
ends on to the pencil by two pins, so that it will not untwist, hold the paper thus twisted on the pencil at right
angles to the surface of a looking-glass : and in the looking-glass you will see its image.   In Diagram I., M repre-

sents the mirror, and on the left hand is shown the twist, on the right hand the image twist. Now take another pencil and another piece of paper, and make a model of what you see in the glass. You will be able to twist this second piece of paper in a spiral round this second pencil so that it is an exact copy of what you see in the glass. Now put the two pencils together end to end, as they would be if the first pencil were to approach the glass until it touched it, meeting its image : you have the real copy of the image instead of the image itself. Now pin together the two ends of the pieces of paper, which are near together, and you have your two Kilkenny cats ready for the fray. To make them fight (remember that the twists—not the paper itself, but the paper twisted— represent the cat), hold firmly and pull the other ends (the tail ends, so to speak), so as to let each twist exercise its nature on the other.

You will see that the two twists mutually annihilate each other. Without your unwrapping the paper the twists both go, and nothing is left of them.

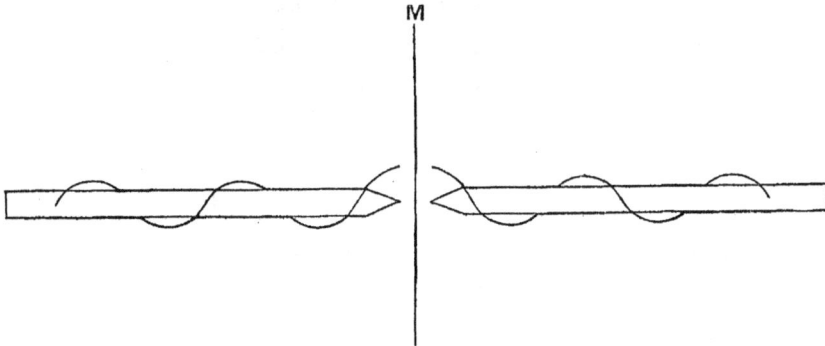

Diagram I.

Now the image of the twist as a real thing was made by us. It did not exist in nature other than as a mere appearance.

But I want you to imagine this process of producing a real image as somehow existing. I want you to lay aside for the present the question of how it could be done, and to conceive twists and image twists.

This is the mechanical conception I wish you to adopt— there are such things as twists. Suppose by some means to every twist there is produced its image twist. These two, the twist and its image, may exist separately ; but suppose that whenever a twist is produced its image twist is also produced, and that these two when put together annihilate each other.

With this conception let us explore the domain of those actions which are called electrical.

When a glass rod is rubbed with silk it becomes excited, its state is different. It manifests many pro- perties, such as that of attracting light bodies, giving off a glow of light, &c. The silk also with which it was rubbed manifests similar properties. It also attracts light bodies, appears to glow in the dark, &c.

And yet there is a difference between the state of excitement of the glass and of the silk. The electricity which is in them is of different kinds. And if the electricity of the silk and of the glass be brought together, all electrical effect disappears ; they become glass and silk in an ordinary condition.

It may seem strange that, if this is so, they should become electrified when rubbed together. Yet this is the case, and must be taken as a fact. It seems to depend partly on the circumstance that glass and silk are not what is called conductors. In a conductor, if one part receives electricity, this electricity at once runs over the whole of the conductor, whether it be an inch long or many feet. And if any part of the conductor be touched by another conductor which is in contact with the earth, every trace of electricity leaves the con- ductor, flowing, as it were, freely out of it.

Now both glass and silk do not let the electricity run

from them so easily. To discharge a glass rod it has practically to be touched in every part. Thus, when by the rubbing with silk electricity is produced on it, it is conceivable that this electricity should be kept to a certain extent, and not combined immediately with the electricity on the silk.

Besides, the same cause—the friction which produced the electricity on the glass, and the other kind of electricity on the silk—would probably prevent their combination as long as it was applied.

Now let us suppose that the electrical charge which the glass has consists in this.

Let us suppose that the particles of the glass on the surface of it are twisted, strained out of their natural position, and twisted.

Let us also suppose that the particles of the silk are twisted too, but let them have the image twist.

Now these two twists, the glass twist and the silk twist, its image, when brought together, will run down. In unwinding each other they will give off a certain amount of energy, which will manifest itself as a spark, make a crackling sound, and so on. But when they unwind each other there is no more tension of the particles.

This does not explain in the least why the glass particles should receive a twist in one direction, the silk particles a twist of the image kind.

But instead of inquiring into this, it is best to see if this supposition is in accordance with other known facts of electrical action ; because, if it is not, we may dismiss it more easily than we could if we had to test it with regard to the very inaccessible question of why some bodies, when rubbed, get electrified in one way, others in another.

When the glass and silk are near together the twist

on the glass and that on the silk are related to each
other as twist and image twist, and there is no action on
surrounding bodies from either of them, as they, so to
speak, satisfy each other.

But when the glass rod and the silk are moved apart,
and brought near other objects, then each of them calls
up on those objects near which it is brought a twist of the
kind which is the image of its own twist.

If the glass rod is brought near a mass of metal,
which we will call a conductor, the following effect is observed :—The part of the conductor near the glass rod
becomes charged with electricity of the silk kind ; the
part of the conductor away from the rod becomes
charged with electricity of the glass kind.

Now let us bring into play the supposition which we
made before.

Let us suppose that there is some process in nature
which, when there is a twist, makes a real image of that
twist come into being. If we assume this process, we
see that, opposite to the silk, on whatever objects are
near it, will be a twist of the glass kind, and opposite
the rod will be a twist of the silk kind. That is to
say, that on the conductor there will be a twist of the
silk kind when the glass rod is brought near it.

But there is nothing to make the particles of the
conductor twist as a whole. The glass rod is not supposed to touch the conductor—it is simply brought near
it, and no actual communication takes place between
them. No force is actually applied to it, nor electricity
communicated to it. Hence, on the whole, the particles
of the conductor will not be twisted. That is to say,
since there is a twist of the glass kind on one end, there
will be at the other end a twist of the image kind—that
is, of the silk kind. And these two twists are like the
two twists on the pencil—if allowed to run together they

will run each other out. So if the conductor were re-moved from the neighbourhood of the glass rod it would be found to be no different from what it was at first. It would not be "charged."

Now this is what actually happens.

Thus we have, firstly, a glass rod with its twist; secondly, a mass of metal with two twists on it—one near the glass rod, and of the image kind ; the other at the other end of the mass of metal, and related to the original twist in the following way. It is the image of its image.

Now it will be found by using a mirror that the image of the image of a twist is the twist itself.

Hence on the other end of the conductor there is a twist of the same kind as that on the glass rod.

And it is obvious that the rod, with its twist, is con-nected with the twist on the conductor nearest to it, and the twist on the other end of the conductor will, by the same arbitrary process which we have assumed as real, call up a twist, the image of itself, on any object near it.

If it is touched by the object with this contrary twist the two will run together, and the conductor will, if it is left free, have only one twist—the silk kind, which has come up opposite the rod.

If now the rod be moved away, the conductor will be twisted as a whole ; that is, all the particles of which it is composed will be slightly twisted with a twist of the silk kind.

In this state it is said to be charged.

Thus the assumption which we have made, that there is some process in nature in virtue of which, opposite to any twist, its image twist is produced as a real thing —this assumption is in harmony with the laws of induction.

Instead of working with a glass rod it is more con-

venient to use a metal rod. Suppose we take a poker
and attach a handle of sealing wax to its middle. See
A B below. This will be easily imagined, and its two
ends, the handle and the black end, will be easily re-
tained in the mind.

If now electricity be communicated from the glass
rod to the poker by touching the two together, what
happens is this : The particles of the glass being twisted
communicate their twist to the particles of the poker.
The twist on the poker is of the same kind as the
twist on the glass rod, and the amount of twisting
which the glass particles had is divided between the
glass rod and the poker. The use of the sealing-wax
handle is to keep the twist from communicating itself
to the body of the person holding it, and, through him,
to the earth. It is found that certain bodies, " non-
conductors," will not communicate the twist and convey
it along ; whereas metallic bodies, and conductors
generally, will communicate this twist at once to great
distances.

We will suppose that a metallic body consists of particles so arranged together that it easily acts as a set of
minute threads or chains of particles which will twist,
each thread or chain twisting as a whole. Thus the
conception which should be formed of a metallic body
conducting electricity along it is this :—Conceive a
bundle of very fine but very rigid wires, each wire twisting separately but with the same kind of twist as all the
others, and each, as it twists, rotating amongst its fellow
threads. If we have a metal rod we can twist it between
the finger and thumb. This is not the kind of twist we
suppose, but that each separate string of particles is thus
twisted, so that each set twisting remains in the same
part of the metal rod—but is turning round in its fixed
position. This is a body conveying an electric current.

If the current will not pass, the set of minute wires must be conceived as held at the far end, and given a twist, starting from the point where the electricity is communicated. Now if a conductor is thus charged and left, it is found that it retains its charge ; to be discharged it must be touched with another conductor. Hence this twist of minute threads differs from a twist of a wire in that the threads cannot untwist of themselves unless other threads come into contact with them to which they can impart the twist. That this should be the case may depend on the fact that the twisting strings are strings of molecules, and the ends of them would thus be connected with other molecules with something of the same tenacity as that with which the strings themselves cohere together, and are unable to unlock themselves from these insulating or untwisting molecules.

Let us consider the state due to these twisting strings of particles.

Place two pennies lying on the table before you, and suppose them to be the sections in which two strings of a conductor are cut across, so that you are looking at two particles, represented by the pennies in the interior of a conductor ; the strings, of which the pennies are sections, come up towards your eye. Now twist the two pennies each in the same direction—say that of the hands of a watch. From the outer edges you can take the motion off; the edges are moving in the same direction. But where the two pennies meet you will see that the edge of each is going in a contrary direction to the other. And if one penny tends to move an object in one way the other tends to move it in the contrary direction. Hence these motions tend to neutralize each other in the interior of a conducting wire.

Having now formed a conception of the state of the

particles in an electrified poker, suppose another poker likewise held by an insulating handle is brought near the first.   Let the pokers be so arranged that the handles both point one way, the black ends another way, and let the second poker be in the same line as the first, with its handle towards the black end of the first.

Now the first poker is charged, it contains electricity, its particles are twisted.   What effect will it have on the second poker?

It is found that the second poker undergoes a certain change, but when it is removed to a distance from the first poker all trace of this change disappears.

On the end nearest the first poker—on the handle—is found silk electricity ; on the end furthest from the first poker—on the black end—is found glass electricity.

A B	C D
+ +	− +

Let A B be one poker, the + representing the charge of glass electricity. Let C D represent the other poker, the — representing the induced silk electricity, the + the glass electricity in it.

Let A be the handle of the first, B its black end.   Let C be the handle of the second, and D its black end.   To explain this let us bring in our imaginary principle. Let us suppose that when a charged body is brought near an uncharged body, but is separated from it by some medium through which electricity cannot pass—let us suppose that by some agency the twist in the charged body calls up an image twist in the body opposite it. Thus, due to the twist in the first poker there will be an image twist in the handle part of the second poker.

But the strings of particles in the second poker are not twisted as a whole; they are twisted in such a way that if they are removed from the first poker, the twist, whatever it be, disappears.

Now this would obviously be effected if we suppose

the same thing to go on in the second poker as took place between the first poker and the second. Let the twist in the handle end C of the second poker be accompanied by the production of an image twist in the black end D. And let us take this as a fair account of our observations. If a body, which as a whole does not undergo a twist, has one part of it twisted, then there will be the image twist in the other part of it. I say that the poker as a whole is not twisted, and all that this means is that if it be removed from the electrical influence it is found to be not charged ; and the idea which we may form is this : the strings of particles are twisted like the two strips of paper round the pencil ; they are twisted so that they will exactly unwind if left alone.

Now of course all these suppositions are merely provisional, and must be dismissed unless seen to be mechanically possible ; but for the present we are trying to see if our assumption will fit in with facts. And our assumption is that there is in nature a power which amongst the molecules produces that as a real thing which in our larger mode of existence only occurs as a simulacrum and appearance. Our looking-glass images are not real, but we suppose that real images are produced amongst the molecules.[1]

We have seen that if we make a certain supposition as to the calling up of an image twist by a twist of molecular matter, then the main facts of electricity are capable of an explanation, which, involving merely the motion of ordinary matter, is far preferable to the idea of there being a mysterious fluid, and more in harmony with our present ideas of electricity.

And yet it is impossible to retain this supposition unless a clear mechanical explanation can be given of

[1] For details, see Appendix III.

how a real image of itself can be called up by the twist which we suppose electricity to be.

We can by intelligent agency produce a twist which is the real image of a given twist. But it would be absurd to suppose amongst the molecules an agency which, acting with prescribed aim, gave in that domain those real simulacra, those evident images, those phantoms with which we in our larger world of masses are for ever mocked.

And yet it would be curious if such an hypothesis were to claim a recognized position in our mental apparatus with which we think about nature.

For in that molecular world, if we imagine it to ourselves, there would be a curious state.

If we consider a twist and its image, they are but the simplest and most rudimentary type of an organism. What holds good of a twist and its image twist would hold good of a more complicated arrangement also. If a bit of structure apparently very unlike a twist, and with manifold parts and differences in it—if such a structure were to meet its image structure, each of them would instantly unwind the other, and what was before a complex and compound whole, opposite to an image of itself, would at once be resolved into a string of formless particles. A flash, a blaze, and all would be over.

To realize what this would mean we must conceive that in our world there were to be for each man somewhere a counter-man, a presentment of himself, a real counterfeit, outwardly fashioned like himself, but with his right hand opposite his original's right hand. Exactly like the image of the man in a mirror.

And then when the man and his counterfeit met, a sudden whirl, a blaze, a little steam, and the two human beings, having mutually unwound each other, leave nothing but a residuum of formless particles.

CHAPTER II.

WHAT physical explanation is possible of this production of a real image ?

First of all we may note that the production of a real image of any disturbance is one of the commonest phenomena.

If a piece of indiarubber lying on the table be pressed downwards with the finger it will move up when the finger is removed. The yielding and the resuming its original form are movement and image movement.

If the disturbance is simply a displacement in one line, then, if the medium in which this displacement is produced is not permanently displaced, but on the whole maintains its equilibrium, there invariably accompanies any displacement its image displacement.

Moreover, to take the simple example of a wave propagated through water—the particles of the water on the whole move about a mean position ; they are not displaced permanently in any one direction ; and, taking the distance from the crest to the hollow of a wave, then from the hollow to the next crest, is the real image of the first part. Thus in the complete movement in the wave measured from crest to crest, there is displacement and its real image.

Thus there seems some consistency about this supposition of an image, about the production of a real image in nature.

But there are two observations which we can make.

Firstly, if it is true in these complicated cases it ought to be true in simpler cases also. That is, if this supposition is in harmony with electrical actions, it ought to fit in with other actions of a simpler kind.

Secondly, a supposition of this kind has no per-

manent value ; it is rather a feeler, by which we trace
out our way in the darkness, than any actual vision
itself.   In default of an actual realization of what the
electrical relations are we can treat them by means of
a supposition.   But we must be ready at any moment
to give up the supposition if it does not harmonize with
the facts.

And in the first case does the idea of a real image
hold good about the simplest possible actions ?

If we push our fist towards a glass the image is that of
a fist moving in the opposite direction.

Now, suppose a pressure exerted on a wall, as, for
instance, a hard stone hitting it.   The wall undergoes a
displacement, but not as a whole—only that part of it
where the stone hits.   And this displacement is followed
by the image displacement, for the wall in the part
where it has been hit and pressed back moves forward,
and by its reaction throws the stone off.

Every case of action and reaction is a case of a motion
and its image motion.

If a bullet strikes the wall and goes with such velocity
that it lodges in it, then the motion of the ball and the
image motion of the wall destroy one another, and the
result is a shattering of the wall in the path of the bullet.

Now in the case of a simple displacement of this kind
there is a rule by which we can form the image displacement.   Take a point on the wall, and about this
point as a centre turn the displacement half way round,
so that it does not come to be itself again, but is opposite to itself.

By this turning, the displacement becomes the image
of itself ; a movement into the wall becomes a movement out from the wall ; and these follow one another if
the wall is not injured.   It should be noticed that the
displacement is moved round this point, using a direction

which is *not* in the displacement itself. The displacement goes straight into the wall. The turning motion, which we suppose, needs another direction than this.

Now suppose, instead of a simple displacement like this, we take a displacement involving two directions, as in the case of a wave disturbance—it will be found that the conditions are just the same. If a wave movement falls on a medium which it does not destroy or move as a whole, the displacement calls up its image displacement. And the image displacement can be found, as before, by twisting the displacement round so as to become opposite to itself—by twisting it half-way round. But in this case, too, a direction must be used which is not used in the displacement itself.

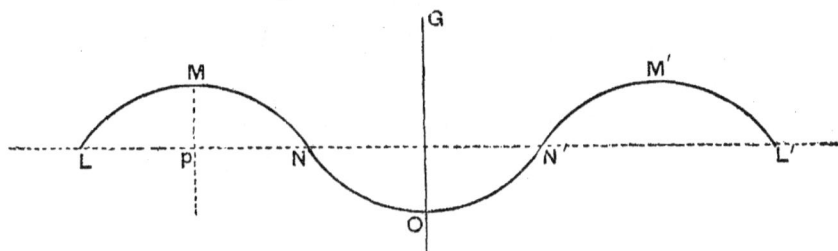

Diagram II.

Let us look at the wave disturbance more closely.

The horizontal central line in Diagram II. will represent the positions which a number of particles occupy when at rest. That is, let us suppose there to be a number of particles lying in a series forming this line.

We can think of the portions of an elastic cord. An indiarubber tube may be taken as an illustration, and made to vibrate by a motion of the hand.

If now one of the particles be deflected from its natural position—suppose it is moved to the position M—then we should have one particle at M out of its place, and all the others in their places.

But this does not happen. If the particle is pulled to

M, the particles near it follow after it, and are also disturbed from their places, though not so much as the particle at M.

We should have a set of particles forming a shape like L M N, only much longer; in fact, the particles all along the cord would be raised.

If the cord is struck suddenly we do have a set arranging themselves like L M N, but only for a limited distance along the cord.

And here we notice a curious thing.

If a set of particles is forced to go like L M N, removed from their position of repose, then at once a set of particles goes like N O N'.

A displacement is accompanied by another displacement which is the opposite of it. And this displacement and opposite displacement travels along the elastic cord.

But the point of view which is the most natural one to regard it from is a little different from this. Let us consider a single point, P. When this is disturbed it moves above its original position to M, and below to the other end of the dotted line. Its complete movement is from one of these extremes to the other. And if we take the complete disturbance as exhibited in all its phases by different points, we ought to look at the portion of the diagram M N O. For here at N we have a point not displaced at all; at M, one displaced to its full extent upwards; at O, one displaced to its full extent downwards. And intermediate particles have intermediate displacements.

Now when a complex displacement of this kind is put into a cord, its image at once springs up. The displacement represented by M N O at once calls up the displacement represented by O N' M', and this condition of displacement and image displacement continues repeating itself till the cord comes to rest.

If the diagram be closely looked at, it will be seen that it exhibits the image relationship twice over. For the movement of the particle P from P to M has its image in the motion of another particle from its place of repose to the position O. The disturbance itself, M N O, consists of displacements and image displacements ; and this disturbance, with its image O N' M', makes the wave from crest to crest.

The " twist " which we consider in these pages is like the wave motion, but with a third component added, so that in the complete motion there is a displacement coming out from the plane of the paper, as well as the displacements in the plane of the paper itself.

And just as the wave displacement produces a real image of itself in a medium which it does not distort as a whole, so there is nothing arbitrary in our assuming that an electric twist calls up the real image of itself in an insulating medium—that is, a medium which it cannot twist as a whole.

If L M N O is a wave motion, then L' M' N' O is its image, as produced by moving it round out of the plane of the paper—Diagram II. If the wave disturbance is moved round in the plane of the paper, the original wave L M N O becomes L' M' N' O—Diagram III.—a shape which bears no resemblance to the transmitted wave.

Consider O N M L to be a bent piece of wire lying on the paper ; if it is moved round O, keeping on the paper, it becomes O N' M' L'. To become like O N' M' L' in Diagram II. it must move up from the paper and down again on the right.

Thus adopting this artificial aid to thought—that a displacement calls up an image displacement—we get the rule that this displacement, the image, can be got from the original displacement by moving the original displacement half-way round, using as the plane in which

the turning is made that plane which is given us by
taking these two directions—the direction in which the
wave is moving, and a direction at right angles to the
directions in which the displacements which form the
wave take place.

Thus, with the wave motion shown, if we take the
direction towards the top of the page to be the up direction, and that from left to right to be the sideways
direction, then out of the paper towards us is the "near"
direction.   So, too, in this case we have to turn the wave
disturbance out of the plane of the paper, and each point
of it, to produce the image, must turn in a circle (going

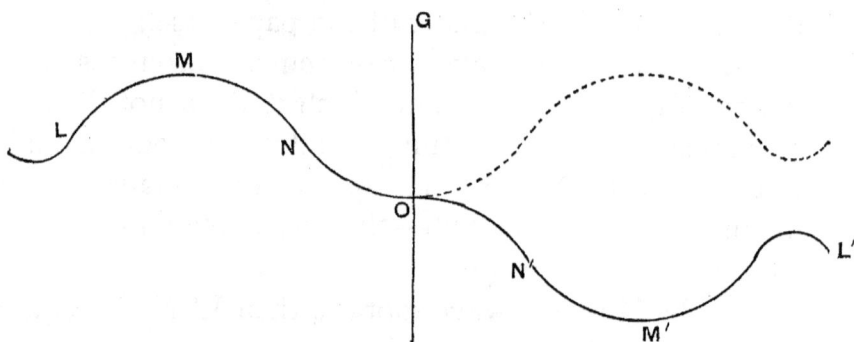

Diagram III.

half-way round it) lying in a plane which has the two
directions near and sideways.   The motions of the particles themselves are in the plane of the paper.   So to
get the image by turning we use a direction — the
"near" direction, which is not involved in the wave
motion itself.

Hence we may state, as a tentative principle, that
when a disturbance takes place in a medium which will
not be disturbed as a whole, then such disturbance is accompanied by a real image of itself; and this real image
of itself is the configuration which would be obtained by
twisting the original disturbance round in a direction not
contained in the original disturbance.

Thus the disturbance O N' M' L' is obtained by twisting the disturbance L M N O round. The direction in which it is twisted is the direction coming out from the plane of the paper.

Now if this plane disturbance is in nature accompanied by its real image, why should not a twist such as takes place in the electric current also be accompanied by its image twist when it impinges on a medium which it cannot twist as a whole—that is, when it comes to an insulator in its path?

The reason, obviously, is that we cannot conceive such an image produced mechanically. And the reason of this can be exhibited thus.

When we had a plane disturbance like L M N O we only used up two dimensions of space, and we have a third coming up from the plane; and this direction enables us to imagine a turning which will alter A B into its image.

But when we have a twist proceeding along an axis, as in the case of electricity, we have no direction left over in space whereby we may conceive the twist turned round.

Now when the displacement itself involves all these directions how will our rule hold?

How shall we get the image displacement? We can find what it is by using a looking-glass; but the same rule which served in previous cases ought to work here also.

We want a direction which is neither up and down, right and left, towards and away.

Now let us adopt a mathematical device, and suppose there is such a direction, and let us call it the X direction, the unknown direction.

Then if we turn the twist round, using this X direction, we shall get the image if our rule is correct. And as a matter of fact, by twisting a figure round in this way, using a direction different from any of the three mentioned above, we do get its image.

Hence the rule we have formed works consistently.

It will be found that if there was another direction so that the spiral disturbance could be turned independently of the directions used up in it, that just as a plane disturbance can be turned into its image disturbance, so the spiral disturbance of electricity could be turned into its image spiral by a simple turning.

In this argument we have not looked at the matter directly, but from the outside. To see it immediately requires us to gain a familiarity with the properties of space with four independent directions, and that would take too long for the present paper. The same conclusion can be arrived at mathematically ; but in these papers as far as possible we avoid symbolism. We want to gain hold of scientific facts in a warm and living way, to unwrap them from conventionalities and formulæ.

Thus if we suppose that in the minute motions which go on about us there is a possibility-of moving in a four-dimensional way, then it is perfectly legitimate to assume that in a medium which cannot be twisted, but which is elastic, a twist calls up a real image twist.

And thus the assumptions which we have made as the basis of an electrical theory are justified on the assumption of a four-dimensional space, are untenable except on that supposition.

The matter is of course perfectly open. The only way is this, by adopting the assumption of a higher space to predict what the actions of the molecules will be, then if a number of predictions are verified the evidence will become strong. And I feel sure that there are some very curious things to be made out here. For my own part the evidence of the reality of four-dimensional space—in the sense in which we say that our space is real—does not rest on the consideration of the molecular movements about which it is not easy to get clear ideas, but on the study of the facts of space. I hardly think that any one who spent a few years in becoming familiar with the facts of space, not by the means of symbolism or reasoning but by pure observation, could doubt that there are really four dimensions.

In noticing the simpler actions and their image actions we find that the real image does not coexist with its original, but rather follows and succeeds it. If we push against a board the board yields, and springs back when we leave off pushing. If the original displacement is permanent as a point pressed against an elastic surface and making the surface yield, then the image of this displacement is potential; it is not actually there, but comes into play as soon as the original displacement is removed.

Now in the electrical actions we have assumed both the original twist and the image twist as concurrently existing.

In certain cases there is no doubt that they are co-existent as when a glass rod is rubbed by silk.

But if the case of the action of a charged poker on an uncharged one be examined it will be found that there is nothing to prove that the image twist comes into existence until the original one is removed.

When the charged poker is brought near the other, the remote end of the second is affected with the same kind of electricity as is on the charged poker.

The appearance is just the same as if a thin wall were exposed to a pressure on one side, and the other side were to bulge out. The displacement is transmitted through the conductor.

It is only when the original charged body is removed that the image charge is found to be in existence on the second conductor. There are some peculiarities, however, which make electrical displacements different in their appearances from ordinary displacements.

No body can be made to move in any direction without imparting an equal motion in an opposite direction to another body—*e.g.*, the motion of a cannon ball is equalled by the recoil of the cannon.

And so no twist can be given to the particles of a body without an image twist being given to other particles.

Now the image displacement or rectilinear motion, in the case of a rectilinear motion, in straightforward movement seems to remain in the place where it was produced. The recoil of the gun carriage produces a strain on its bearings and friction, which produce heat, which gradually dissipates.

But the image displacement, in the case of electricity, seems to have a marvellous facility for running through the earth and meeting the original displacement. An indefinitely long line of action seems in electricity to take the place of a simple point. Our ordinary mechanical forces are located in centres, or points of action. In electricity the line seems to take the place of the point. Where the ordinary engineer deals with points the electrical engineer deals with lines.

CHAPTER III.

THE ÆTHER.

THERE are some expressions which, being somewhat vaguely used, are apt to cause confusion in the mind of those who read or hear about higher space.

And perhaps the most mischievous is the expression, a curvature of space. Now of space as it is generally used, in its accepted significance, there can be no curvature. For space means a system of positions extending uniformly in the number of dimensions we choose to fix upon.

If we take the straight line as our space, we may call it 1 space ; then the set of positions follow one on after the other without bending. If the line is bent it becomes a *line*, not a straight line. It should not be called 1 space, but a thing in 2 space. That is, it is a bent line in a plane.

A being who was on the line might not perceive the fact of this bending, and it might not affect the measurements he made. But if the line ran into itself again, and he found that he was moving on what we should call a circle, this would in no way affect his idea of space. He would recognize that what he called space, namely, his line, was not space, but a curved thing in 2 space.

Similarly, taking a plane—this is by definition not curved in any way, known or unknown, and it can only be conceived to be bent by ampler space being conceived, and its being imagined as having force applied to it so as to become a bent thing in this ampler space. In this case the term " plane " is not the correct name.

And so about our three-dimensional space ; we cannot be robbed of that idea, although it might conceivably be proved that our earth and our whole universe were on a curved thing in 4 space.

We will then keep the term " space " for the ordinary conception ; and call it 1, 2, 3, 4 space, according to the number of supposed independent directions.

A curved line or surface or solid we will call a 1, 2, or 3 thing, according to the number of dimensions in it.

A straight line is a 1 thing possible in 1 space. A circle is a 1 thing possible in 2 space. At any point of it a being in it is limited to motion in one direction, while the circle itself involves two dimensions. The surface of a sphere is a 2 thing possible in 3 space. The rind of an orange, or the orange itself, is a 3 thing possible in 3 space.

It will be observed that the surface of the sphere, although only a 2 thing, involves the conception of 3 space, and cannot be understood without the use of the idea of 3 space. It is a 2 thing because at any point of the surface a being can only move in two independent directions. A crooked line drawn on the surface of a sphere is a 1 thing in a 2 thing in 3 space.

Another very common misconception is occasioned by the use of a figure of this kind ⚇ to represent a " knot " in 2 space.

It obviously corresponds in 2 space to an iron rod welded together at the crossing place of the loop, so that it is indistinguishable which is the one free end, which the other. At the crossing point the two lines represented by the two ink marks must be absolutely one and the same.

If one line be supposed to go over the other, by however small a distance, it would leave the plane. It would suddenly become invisible to the creature in the plane, and it would appear again at the other side of the line it crossed as if it came from nowhere.

It would be as extraordinary a sight as if we saw a pole going up to a brick wall, then beyond the brick wall the rest of the pole appearing—not going through the brick wall, nor coming round it—but somehow appearing; part of the same pole moving when it moved, obviously connected with it, and yet with no joining part which we could possibly discover.

Again, it sometimes appears to be thought that the fourth dimension is in some way different from the three which we know. But there is nothing mysterious at all about it. It is just an ordinary dimension tilted up in some way, which with our bodily organs we cannot point to. But if it is bent down it will be just like any ordinary dimension : a line which went up into the fourth dimension one inch will, when bent down, lie an inch in any known direction we like to point out. Only if this line in the fourth dimension be supposed to be connected rigidly with any rigid body, one of the directions in that rigid body must point away in the fourth dimension when the line that was in the fourth comes into a 3 space direction.

If the reader will refer back to the paper on the plane world he will find a description of the means by which a being there might know that he was in a limited world, and that his conception of space was not of what was really the whole of space, but of the limited portion of it to which he was confined by his manner of being.

The test by which such a being could discover his limitation was this. He found two things, each consisting of a multitude of parts—two triangles ; and the relationship of the parts of the one was the same as the relationship of the parts of the other. For every point in the one there was a corresponding point in the other. For every pair of points in the one there was a corresponding pair of points in the other. In fact, considered as systems made up of mutually related parts, each was the same as the other.

Yet he could not make these two triangles coincide.

Now this impossibility of bringing together two things which he felt were really alike was the sign to him of his limitation ; and by reflecting on the similar appearance which would present itself to a being limited to a straight line—by thinking of two systems of points which were really identical, and which he could make coincide, but which a line being could not make coincide, he would be led to conclude that he in his turn was subject to a limitation.

Now is there any object which we know which, considered as a whole consisting of parts, is exactly like another whole, the two having all their parts similarly arranged, so as to form in themselves two identical systems, and yet the one incapable of being made to coincide with the other, even in thought ?

Let us look at our two hands.

They are (except for accidental variations) exactly alike.   And yet they cannot be made to coincide.

And here, if we reflect on it, is the sign to us that we are limited in our notions of space—that we are really in a four-dimensional world.

Watching a ship as it recedes from the shore we see that it becomes hull down before it vanishes, and know that the earth is round.   And no less certainly do our two hands, in their curious likeness and yet difference, afford to us a perpetual proof of our limitation, and indicate a larger world.

This sign really tells us more than the mere fact of our limitation : it tells us where to look for the possibility of four-dimensional movements.   It tells us that movements of any degree of magnitude relative to us are not possible in the fourth dimension.   It tells us to look for four-dimensional movements in the minute particles of matter, not in the movements of masses of about our own size.

The task before us is difficult. We have to make up from the outside what the appearances of a higher space existence are to us in our space, and then we have to look at the facts of nature and see if they correspond to these appearances.

Let us take a few isolated points and look at them patiently.

To a being standing on the rim of a plane world a straight line absolutely shuts out the prospect before him. If the straight line is infinite it cuts his world in two ; he can never hope to get beyond it.

It is to him what an infinite plane would be to us, stretching impassably in front of us, cutting us off from all that lies on the other side.

But we know that a point can move round this line. It can revolve round it by going out of the plane, and coming down again into the plane on the other side of the line.

This movement would be inconceivable to a plane being ; for he can only conceive it possible to get to the other side of the line by going to the end of it and coming back along the other side of the line.

Now take a piece of paper and put a dot right in the middle and suppose that it has no means of passing through the paper. We can only conceive the dot getting to the other side of the paper by passing round the edge and coming back again to the position underneath where it was.

But by a four-dimensional movement it can slip round the paper without going to the edge.

A set of words may help. In a plane a body rotates round a point—rotation takes place round a point. In space rotation is always round a line—the axis. In four-dimensional space rotation takes place round a plane.

To take a farther consideration of this point—a plane being can see one side or the opposite of a straight line. He can only see it in one direction or in the reverse direction. But we can look at a straight line from a direction at right angles to that in which a plane being looks at it. We can look at a straight line from points which go all round it.

Similarly, a being in four-dimensional space can look at a plane from a direction at right angles to that in which we look at it. If we try to think of this we shall imagine ourselves looking at the thin edge. But this is not what a four-dimensional being would mean. He would see the plane exactly as we see it, but it would be from a direction at right angles to that in which we look.

In working with four-dimensional models it is a curious sensation until we become used to it—that of looking at a plane at one time, and then looking at it again ; and, although it seems just the same—as square in front of us as before—realizing that we are looking at it from a direction at right angles to that of our former view.

And in four dimensions a point which is quite close to a plane can revolve round it without passing through it, thus presenting to us the appearance of vibrating across the plane, but not passing through it.

The appearance is as wonderful to us as it would be for a plane being to see a point which was in front of a line quickly passing behind it without having gone round the end. Such a point would appear to the plane being to vibrate across his line without passing through it.

Now if we stand in front of a mirror we see the image of ourselves. If we were to go round the mirror and take behind it the position which our image seemed to

occupy, we should not be able to make ourselves coincide with it.   In the mirror opposite to our left hand
is the image of our left hand ; but if we passed round,

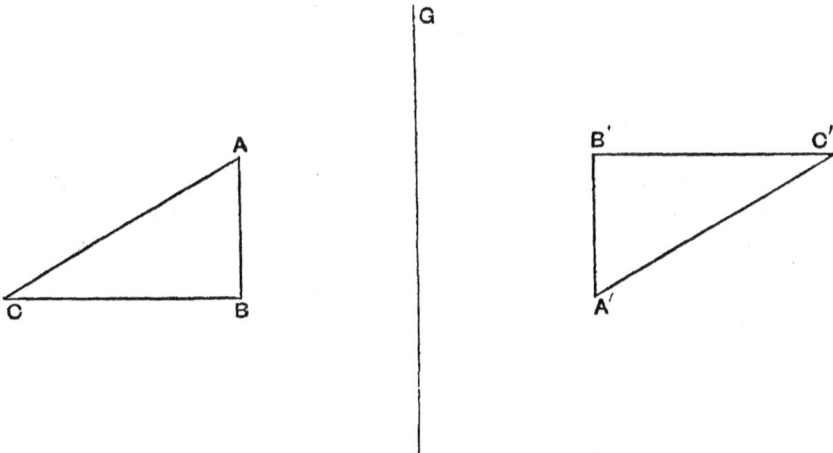

Diagram IV.

our right hand would be in the place in which we
imagined we saw the image of our left hand.   And thus
we cannot make ourselves coincide with our image.   But
by a rotation in four-dimensional space we could put

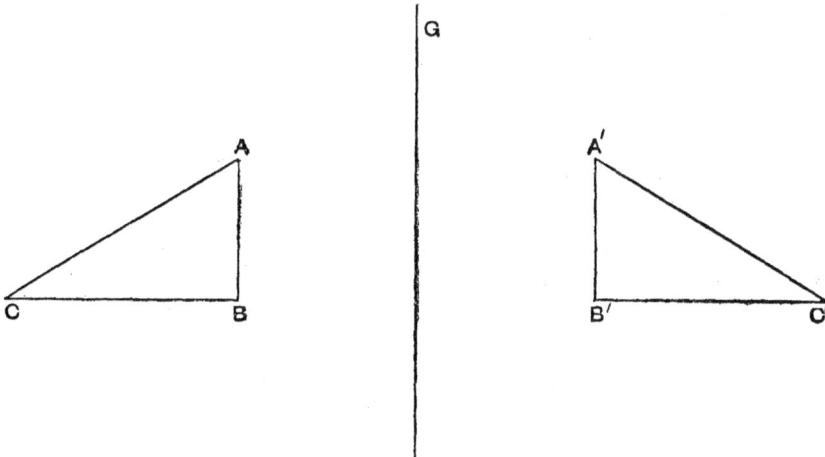

Diagram V.

ourselves so as exactly to coincide with our image.
This can be seen by referring to the case of the straight
line, Diagram IV.

Let A B C be a triangle, and G a line. If A B C moves round the end of the line, it can take up the position A′ B′ C′; but it cannot anyhow be made to take the position shown in Diagram V., A′ B′ C′.

But if we move the triangle A B C out of the plane round the line G as axis, it will, in the course of its twisting round this axis, come into the position A′ B′ C′. It will come into this position when it has twisted half-way round. The point A, for instance, twists round in a circle lying in a plane which contains the direction A to A′, and the direction at right angles to the paper. Twisting half-way round in this circle, it becomes A′, and so on for the other points. Now a being who did not know what a direction was which lay out of the plane would not be able to conceive this twisting and turning movement. It would be as impossible for him to conceive the triangle A B C turned into the triangle A′ B′ C′, as it would be for us to suppose ourselves turned into the looking-glass image of ourselves by a simple twisting.

Yet just as a thing inconceivable to the plane creature can be done, so we could be twisted round and turned into our image. But this only holds theoretically; our relation to the æther is such that we cannot be so turned, or any bodies of a magnitude appreciable to our senses.

If we consider the case of a being limited to a plane, we see that he would have two directions marked out for him at every point of the rim of matter on which he must be conceived as standing. This is up and down, and forwards and backwards—the up being away from the attracting mass on which he is.

Now, if he were to realize that he was in three-dimensional space, but confined to a plane surface in it, his first conclusion would be that there was a new direction starting from every point of matter, and that this new direction was not one of those which he knew. This new direction he could not represent in terms of the directions with which he was familiar, and he would have to invent new terms for it.

And so we, when we conceive that from every particle of matter there is a new direction not connected with any of those which we know, but independent of all the paths we can draw in space, and at right angles to them all—we also must invent a new name for this new direction. And let us suppose a force acting in a definite way in this new direction. Let there be a force like gravitation. If there is such a direction, there will probably be a force acting in it; for in every known direction we find forces of some kind or another acting. Let us call away from this force by the Greek word ana, and towards the centre of this force kata. Then from every point in addition to the directions up and down, right and left, away from and towards us, is the new direction ana and kata.

Now we must suppose something to prevent matter passing off in the direction kata. We must suppose something touching it at every point, and, like it, indefinitely extended in three dimensions.

But we need not suppose it—this unknown—to be infinitely extended in the new direction ana and kata. If matter is to move freely, it must be on the surface of this substratum. And when the word surface is used it does not mean surface in the sense that a table top is a surface ; it is not a plane surface, but a solid space surface. If from every point of a material body a new direction goes off, the matter which fills up the space produced by the solid moving in this new direction, will have the solid it started from as its surface, and will be to it as a solid cube is to the square which bounds it on the top.

Now this body which extends thus, bearing all solid portions of matter in contact with its surface by every point of them, may be thick in the kata direction or thin.

If it is thick, then the influence of any point streaming out in radiant lines will pass as in all space directions, so out also in this new direction.

And then if its influence spreads out in this new direction, its effect on any particle near it will diminish as the cube of the distance ; for, besides filling all space, it will have also to fill space extended in this new direction.

But we know that the influence proceeding from a particle does not diminish as the cube of the distance, but as the square of the distance.

Hence the body which, touching all solid bodies by every point in them, and supports them extending itself in the kata direction—this body is not thick in this direction, but thin. It is so thin that over distances which we can measure the influence proceeding from a body is not lost by spreading in this new kind of depth.

Thus the supporting body resembles, as far as we know it, a portion of a vast bubble. But moving on the surface of this bubble we can pass up and down, near and far, right and left, without leaving the surface of the bubble. The direction in which it is thin is in a direction which we do not know, in which we cannot move. But although we cannot make any movements which we can observe with our eyes in this direction, still the thin film —thin though infinitely extended in any way which we can measure—this thin film vibrates and quivers in this new direction, and the effects of its trembling and quivering are visible in the results of molecular motion. It only affects matter by its movement in directions at right angles to any paths which we can point to or observe, and these movements are minute; but still they are incessant, all-pervading, and the cause of movements of matter. It is smooth—so smooth that it hinders not at all the gliding of our earth in its onward path. Hence it does not transmit a direct pull or push in any direction from one particle to another; but by the twistings and vibrations of the material particles it is affected, and conveys from one to another these movements. Yet to bear up all matter, and thus hold it on its vast solid surface, it must be extremely rigid and unshatterable; and hence it cannot be permanently altered or twisted by any force proceeding from matter; but receiving from matter any push or twist, it is impressed with it for some distance; then, reasserting itself, it produces an image displacement or twist, and this image it transfers to the particles of matter which it touches.

Sometimes, as when light comes from the sun, this displacement and image is repeated and repeated innumerable times before at last we, receiving it, become aware of the origin of the disturbance.

But the properties and powers of this solid sheet—this film quivering and trembling, yet infinite and solid—are too many to begin to enumerate. The æther is more solid than the vastest mountain chains, yet thinner than a leaf; undestroyed by the fiercest heat of any furnace, for the heat of the furnace is but its shaking and quivering; bearing all the heavenly bodies on it, and conveying their influence to all regions of what we call space.

And by some mysterious action it calls up magnetism from electricity; by its different movements it gives the different kinds of light their being.

Of itself untrammelled and unclogged by matter, it vibrates and shakes with the speed and rapidity of the vibrations of light. But when matter lies on it—when air,

even in its rarest condition, lies on it—its proper movement is damped and some of its quick shakings that are
light, slow down to the obscure vibrations of heat. Thus of
itself it will not take up the vibration of a hot body, but
selects only those orbs which are glowing with radiant
light wherefrom to take its thrilling messages.   But
when matter lies on it, it takes obediently the less
vivacious movements of terrestrial fires.

A being able to lay hold of the æther by any means
would, unless he were instantly lost from amongst us by
his staying still while the earth dashes on—he would be
able to pass in any space direction in our world.   He
would not need to climb by stairs, nor to pass along
resting on the ground.

And such a being, even as thin as ourselves, and as
limited, if not even in physical powers, but merely in
thought he became aware of his true relation to the
æther, he would see all things differently.

From all shapes would fall that limitation of thought
which makes us see them differently to what they are ;
and in largeness and liberty of possible movement his
mind would travel where ours but creeps, and soar and
extend where ours journeys and diverges.

It is impossible in contemplating the rudiments of
four-dimensional existence to prevent a sense of largeness
and liberty penetrating even through the profoundness
of our ignorance.

Whether we shall find beings other than ourselves,
when we have explored this larger space, cannot be said.

But there is a path which holds out a more distinct
promise.

When the conditions of life on a plane are realized it
becomes evident that much of that which is to us merely
natural—obvious from the very conditions of our life—
could only be attained by beings on the plane as the

result of artificial contrivances and modifications of their natural tendencies. In their progress and development they would, as it were, represent on the plane the features of the normal and undeveloped life of three-dimensional beings, and they would attain, as a result of moral labour and energy, a position which children in our higher life are born to without trouble or thought.

And so we in our advancing civilization may to the eyes of some higher beings represent in our arrangements and institutions an approach to the simplest matters of fact in their existence. We are separated from such a view by our bodily conditions, but we are not to be prevented from taking it with our minds.

By building up the conception of higher space, by framing the mechanics of such a higher world, we may arrive at a fairly accurate knowledge of the conditions of life in it.

And then, with that element in our thought, with the reasoned-out characteristics present to our minds of what life on a higher physical basis would be, we may be able to judge amidst conflicting tendencies with more certainty and calmness.

In one of the following papers of this series an account will be given of some of the facts which we can discern about the machinery and appliances of four-dimensional beings.

But the work of real discernment belongs to those who will from childhood be brought up to the conception of higher space.

APPENDIX I.

SUPPOSITION can be made with regard to the æther which renders clearer an idea often found in literature.

This idea is that of the freedom of the will. If the will is free, then it must affect the world so as to determine chains of actions about which the mechanical laws hold true. We know that these mechanical laws are invariably true. Hence, if the will is an independent cause, it must act so that its deeds produce to us the appearance of a set of events determined by our known laws of cause and effect. The idea of the freedom of the will is intimately connected with the assertion that apparent importance, command of power, greatness and estimation, are outside considerations, not affecting the real importance and value of any human agent. These ideas can easily be represented using the idea of the æther as here given.

For suppose the æther, instead of being perfectly smooth, to be corrugated, and to have all manner of definite marks and furrows. Then the earth, coming in its course round the sun on this corrugated surface, would behave exactly like the phonograph behaves.

In the case of the phonograph the indented metal sheet is moved past the metal point attached to the membrane. In the case of the earth it is the indented æther which remains still while the material earth slips along it. Corresponding to each of the marks in the æther there would be a movement of matter, and the consistency and laws of the movements of matter would depend on the predetermined disposition of the furrows and indentations of the solid surface along which it slips.

The sun, too, moving along the æther, would receive its extreme energy of vibration from the particular region along which it moved, and the furrows of the intervening distance give the phenomena actually observed of our relationship to the sun and other heavenly bodies.

Thus matter may be entirely passive, and the history of nations, stories of kings, down to the smallest details in the life of individuals, be phonographed out according to predetermined marks in the æther. In that case a man would, as to his material body, correspond to certain portions of matter; as to his actions and thoughts he would be a complicated set of furrows in the æther.

Now what the man is in himself may be left undetermined; but he would be more intimately connected with the æther than with the matter of his body. And we may suppose that the æther itself is capable of movement and alteration; that it moulds itself into new furrows and marks.

Thus the old woman smoking a pipe by the wayside years ago, and whom I somehow so often remember, is not much different from me—we are both corrugations of the same æther.

Now our consciousness is limited to our bodily surroundings. Yet it may be supposed that in an action of our wills we, whatever we are (and for the present let us suppose that we are a part of the æther), we may be altering these corrugations of the æther. A single act of our wills, when we really do act, may be a universal affair with quite infinite relations. Thus it may be the immediate presentation to us of an alteration proceeding from us of all that set of corrugations which represents our future life ; it may be the whole disposition and lie of events, which are prepared for the earth to phonograph out, being differently disposed.   And it evidently is quite independent of the particular furrows in which such alteration first occurs.   That long strip of æther which is a very humble individual may, by an act of self-configuration, affect the neighbouring long strips and produce great changes.   At any rate the intrinsic value of the will is quite independent of the kind of furrows along which any material human body is proceeding.

APPENDIX II.

IT is a good plan in fixing our attention to give definite names to the directions of space.   Let U stand for up.   Then the up direction we will call the U direction, or simply U.

Then sideways, from left to right, we will call V, so that moving in the V direction, or moving V, means moving to the right hand.

Then the away direction we will call W, so that a motion which goes away from us we call a W motion, and its direction we call W.

Then any other direction which we suppose independent of these we will call the X direction.   Now the simple push or displacement takes place in direction V, or left to right.   It is turned into its image by turning in the plane U V—*i.e.*, the plane of the paper.

The wave motion takes up the directions U V, and it can be turned into its image by a turning in the plane W V—*i.e.*, by turning out of the paper, as if the paper

were folded over about the dotted line. Then finally
the twisting motion takes up the directions U V W, and
can be turned into its image by being turned in the plane
V X. That is, if each point is turned half-way round in
this plane it becomes the corresponding point in the
image twist. Thus on the supposition of the preceding
pages, if a positively electrified particle could be turned
in 4 space, it would become a negatively electrified
particle.

APPENDIX III.

IT remains now to examine if the supposition that the
particles of a wire are twisting in strings fits in with
observed facts of electricity.

And firstly, if the particles are twisting in this manner,
it is only reasonable to suppose that they would take up
a little more room than they did when not subject to
this movement—that is, the wire would become a little
thicker. But its volume remaining the same, if it becomes
thicker it must compensate for this thickening by becoming shorter. And it is found that a wire through
which an electric current is sent tends to become shorter
when the current comes into it.

Again, suppose a wire through which a current has
been sent suddenly isolated. It has a twist in it, and
will keep this twist. But if it is connected up with any
other wire forming a complete circuit through which it
can untwist itself, it will probably do so, and in untwisting would very likely overshoot the mark and
become twisted in the opposite direction. Thus it would
make a series of twists, each less than the last before
becoming quiescent. And it is observed that a wire if
so isolated does produce a rapidly alternating series of

very minute currents before it comes to rest ; just as if it were untwisting itself and overshot the mark each way many times before the electrical state has altogether disappeared.

The question now comes before us, How is it that a wire gets twisted ?   Through what agency is a current of electricity urged through a wire, or a twist put into it ?

This is often done by means of an electrical battery. We will take a simple instance.

Suppose a dish of sulphuric acid, and a bit of carbon and a piece of zinc put into it.   Then the carbon and the zinc are connected outside the liquid by a wire. Along this wire electricity will pass.   Now the twist put into the wire must come from somewhere.   And it is found that the sulphuric acid, which is a very lively compound, and contains a great deal of energy, becomes quieted down, and is quite different after the battery has finished working.   On examination afterwards it is found to consist of sulphate of zinc.

Sulphuric acid can be looked upon as consisting of two bodies—hydrogen and a sulphur and oxygen compound. This sulphur and oxygen compound is called SO_4.   Now the SO_4 comes to the zinc, and with zinc forms quite a dead compound, with little energy in it, called zinc sulphate, or $Zn\,SO_4$.   The hydrogen, on the other hand, comes off at the carbon in an energetic state.

Hence evidently the SO_4 has given up its energy, the hydrogen has not.   So the twist in the wire probably comes from the SO_4 ; and thus the twist is started at the zinc end, and runs round the wire from zinc to carbon.

At the same time we may suppose that an image twist, starting also from the zinc, runs through the fluid of the battery and then along the wire, till meeting the twist the two mutually unwind each other.

Thus the battery will be as if one had a loop of thread, and at one point twisted it between one's finger and thumb. Twist and image twist, starting from this point, unwind each other on the opposite part of the loop. And if the loop is not joined, but the threads are held, each will become twisted with increasing tension till they can twist no longer. The objects which hold the ends of the thread, and prevent them twisting, represent insulators.

It is found that when a strong current of electricity passes through some water which has had a little sulphuric acid added to it, two effects take place.

In the first place some of the current passes through as through a wire. In the next place a part of the current is used up in producing an effect on the water. It splits the water up into two parts, each of them containing very much more energy than the water. One part is called hydrogen, and comes off at the wire which comes from the zinc, which we will call the zinc wire. The other part of the water comes off at the wire coming from the carbon, or at the carbon wire, and is called oxygen.

Let us now suppose that the twist of the zinc wire calls up in the molecule of water next to it an image twist. If it could pass on its twist at once, the water would form an ordinary conductor ; but the water is not a conductor. Hence we suppose the same relation to hold good between the end of the zinc wire and the water molecules as between the zinc wire and any other body to which the twist cannot be communicated.

Now in the part of the molecule nearest the zinc wire an image twist is called up. And hence the molecule, being unable to twist as a whole, in the end of it away from the zinc wire a twist is produced. Thus the water molecule is strained into image twist and twist. Now

14

let us suppose that by a powerful current it is wrenched in two. It is separated into a part having an image twist "hydrogen," which comes off at the zinc wire, and into a part with the twist "oxygen."

But this part with the twist calls up an image twist in the molecule next to it, wrenches it in two. Thus the oxygen of the first molecule separates up the next molecule into hydrogen and oxygen. The oxygen has a twist, the hydrogen an image twist. These twists run each other out, and leave an oxygen part free.

This oxygen part does the same to the next molecule, and so this action is transmitted through the whole body of the water till the carbon wire is reached—when, the oxygen part finding no other molecule to wrench asunder, is left isolated, and comes off in the form of gas.

Thus we see that oxygen and hydrogen would be bodies having in them twist and image twist—that is, that they would have an active rotation each of them; but the rotation would be different in the two cases, and such that if put together they would run each other out: the light and heat produced by the union of the two being probably the exhibition of the effects of this running out.

If we adopt the supposition, which seems most in accordance with facts, that there are in water two different elements occurring in distinct particles, the one called oxygen, the other hydrogen; and if, moreover, we suppose that these particles are perpetually changing places, and that each oxygen particle is sometimes linked with this hydrogen particle, sometimes with that, then it is obvious that the oxygen and the hydrogen in the water are in such a state that, if collected together separately, they would form liquid oxygen and liquid hydrogen; and the effect of the electric twist is to give them those active image rotations, or strains, which make them take the gaseous form, and assume that peculiar relation to each other which exhibits itself so strikingly in combustion.

With regard to magnetism, the same phenomenon of a particular state or disturbance of matter and its image state or disturbance is very strikingly obvious.

For take the case of a magnet. By the influence of an electric current passing round it, it can be turned into a magnet with opposite poles. That is to say, the small particles of the iron have been so shifted that, whatever their disposition was in the first case, they have now the reverse disposition. If we suppose the small particles to be magnets like the whole magnet, and all to have their north poles pointing in one direction, then after the action of the current they have their north poles pointing in the opposite direction. But they have not turned in space, for, if they were to turn, each must turn about some axis. But if there was some axis then, with regard to this axis, the magnetic influence would have a definite relation ; the turning of the particles would take place in a certain plane, and there would be a certain plane in the magnet which would have special properties.

But a magnet is perfectly symmetrical in all its properties round its axis. The magnet which has had its poles reversed is, as an arrangement, the image of itself in its first condition. In the solid mass of iron which forms the magnet, by the action of electricity, a particular arrangement and its real image are alternately produced.

There are some very important electrical phenomena which have been left out of consideration altogether—namely, the repulsions and attractions exercised by electrified bodies.

Adopting the conceptions here laid down with regard to electricity—that the two kinds are in the relation of twist and image twist—we find that certain conclusions force themselves upon us.

A positively electrified body attracts a negatively electrified body.

A positively electrified body repels a positively electrified body.

Or, as it is put in a shorter form, one kind of electricity attracts the opposite kind, and repels the same kind.

Now, if our theory is true, a twist ought to attract its image twist, and repel a twist like itself.

And as far as can be observed it is always a fact that a movement of any kind taking place in a medium does attract its image movement, and repel a movement like itself.

Some very instructive experiments have been made with bodies suspended in water, and caused to pulsate or twist. It would be found, on referring to the details of these experiments, that if two spheres are pulsating or throbbing, so that the movements of the one *are* at any instant what the movements of the other would *seem like*, if looked at in a mirror, then these two spheres will attract each other. If the one is a real copy of the other, then they repel each other. And this law holds good not only for throbbing movements, but also for twisting movements.

If now we supposed that what held good for movements held good for tensions of the same nature as the movements, these results would be in exact accordance with our suppositions. If a twisting movement attracts its image twisting movement, will a twist attract its image twist by means of its effect on the medium in which it is, and on which it exerts tension? This point must be left undecided.

Casting Out the Self.

T HE words which I have chosen as the title of this paper are the expression for a process which has been asserted to be one that occurs alike in our mental and in our moral life. It has so happened that in certain of my own inquiries I have applied this process ; and the details may be of interest. But I must warn the reader not to expect any wide views on life, or far-reaching thoughts, or any of the warmth of human affairs. What I think about is Space ; and it is the application of the principle of casting out the self in attaining a knowledge of Space about which I have something to say.

And, firstly, as a bit of absolute human experience is never without value, but that which we make up is often so, I may as well cast the fear of ridicule aside and enable the reader to take in, in a few lines, the exact commencement of my inquiry.

The beginning of it was this. I gradually came to find that I had no knowledge worth calling by that name, and that I had never thoroughly understood anything which I had heard. I will not go into the matter further ; simply this was what I found, and at a time when I had finished the years set apart for acquiring knowledge, and was far removed from contact with learned men. I could not take up my education again, but although I regretted my lost opportunities I determined to know something. With this view I tried to acquire knowledge in various ways, but in all of them knowledge was too impalpable for me to get hold of it. And I would earnestly urge all students to make haste in acquiring real knowledge while they are in the way with those that can impart it ; and not rush on too quickly, thinking that they can get knowledge afterwards. For out in the world knowledge is hard to find.

At length I came to find that the only thing I could know was of this kind. If, for instance, there were several people in a room, I could not know them themselves, for they were too infinitely complicated for my mind to grasp ; but I could know if they were at right or left hand of one another, close together, or far apart. And the same of, to take another instance, botanical specimens in a book. I could not grasp the specimens— each was too infinitely complicated, and each part too infinitely complex—but I could tell which specimen was next which.

Accordingly, being desirous to learn something thoroughly, and since, in the arrangement of any different objects, there was such a lot of ignorance introduced by the objects being different—each bringing in its own ignorance and feeling of bewilderment—I determined to learn an arrangement of a number of objects as much alike as possible.

Accordingly I took a number of cubes, which were as simple objects as I could get, arranged them in a large block, and proceeded to learn how they were placed with regard to each other. In order to learn them I gave each of them a name. The name meant the particular cube in the particular position.

Thus, taking any three names, I could say, about the three cubes denoted, how they were placed with regard to one another : one, say, would be straight above the first with four intervening, the third would touch the second on the right hand, or some similar arrangement.

Now in this way I got what I conceived to be knowledge. It was of no use or beauty apparently, but I had no reason to use it or to show it.

It is about this bit of knowledge that I want to speak now—a block of cubes, and the cubes are known each one where it is.

Sometimes I have been tempted to call this absolute knowledge, but have been reminded that I did not know the cube itself. Against this I have argued. But in argument we say many things which we do not understand, and my conclusion is, on the whole, that the objection is well founded. Still, if not knowledge absolute, the knowledge of this block approaches more nearly to knowledge absolute than any other with which I am acquainted, because each cube is the same as its neighbour, and instead of an arrangement of all sorts of diverse ignorances we have only one kind of ignorance —that of the cube. Each of the cubes was an inch each way, and I learnt a cubic yard of them. That is to say, when the name of any cube was said, I could tell at once those which it lay next to ; and if a set of names were said, I could tell at once what shape composed of cubes was denoted. There were 216 primary names, and these, taken in pairs, were enough to name the cubic yard.

For the practical purpose of this paper, however, it will suffice if the reader will imagine a block of twenty-seven cubes, forming a larger cube, each cube being denoted by a name (see Diagram I. below). Then it is evident that two names mean a certain arrangement consisting of two cubes in definite places with regard to one another—three names denote three cubes, and so on.

And I would ask the reader not to mind taking a little trouble at this point, and to look at the diagram for a little while. If there is anything about which we can form perfectly clear ideas, it is a little heap of cubes. And if the reader will simply look at them for a little space of time, he will realize clearly every word of what I have to say ; for I am going to talk about nothing else than this little block of cubes.

Thus, looking at the cube with the figure 1 upon it, this numeral will serve for the name of the cube, and similarly the number written on every cube will serve for its name. So if I say cubes 1 and 2, I mean the two which lie next to each other, as shown in the diagram ; and the numbers 1, 4, 7, denote three cubes standing

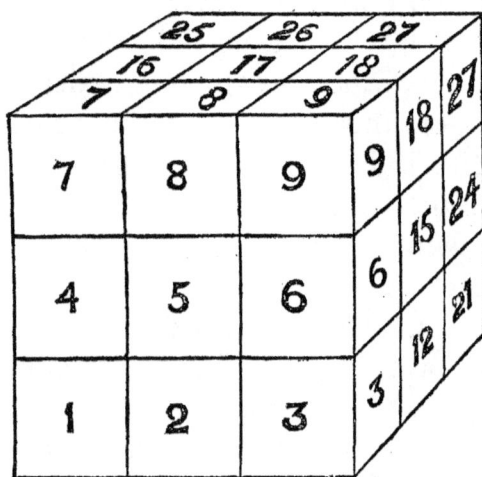

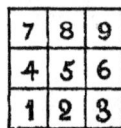

Diagram I, a block consisting of 27 cubes.

Diagram II.

7	8	9
4	5	6
1	2	3

above each other. If I say cubes 1 and 10, I mean the first cube and one behind it hidden by it in the diagram.

Now this is the bit of knowledge on which I propose to demonstrate the process of casting out the self. It is not a high form of knowledge, but it is a bit of know-ledge with as little ignorance in it as we can have ; and just as it is permitted a worm or reptile to live and breathe, so on this rudimentary form of knowledge we may be able to demonstrate the functions of the mind.

And first of all, when I had learnt the cubes, I

found that I invariably associated some with the idea
of being above others. When two names were said, I
had the idea of a direction of up and down. But with
regard to the cubes themselves, there was no absolute
direction of up or down. I only conceive of an up
and down in virtue of being on the earth's surface, and
because of the frequent experience of weight. Now this
condition affecting myself I found was present in my
knowledge of the cubes. When certain of the names
were said, I conceived of a figure having an upper part
and a lower part. Now, considered as a set of cubes
related to one another and not to me, the block had
nothing to do with up and down. As long ago as
Ptolemy, men have known that there is no such thing
as an absolute up and an absolute down. And yet I
found that in my knowledge of the set of cubes there
was firmly embedded this absolute up and this absolute
down. Here, then, was an element arising from the particular conditions under which I was placed, and the
next step after recognizing it was to cast it out. This
was easily done. The block had to be turned upside
down and learnt over again with the cubes all in their
new positions. It was, I found, quite necessary to learn
them all over again, for, if not, I found that I simply went
over them mentally the way first learnt, and then about
any particular one made the alteration required, by a
rule. Unless they were learnt all over again the new
knowledge of them was a mere external and simulated
affair, and the up and down would be cast out in name,
not in reality. It would be a curious kind of knowing,
indeed, if one had to reflect what one knew and then,
to get the facts, say the opposite.

It may seem as if, when the cubes were known in an
upright position, they would be easily imagined in an
inverted position. But practice shows that this is very

far from being the case. It requires considerable mental effort to determine the alterations in position, and to get an immediate knowledge requires a considerable time.

It may seem as if it were a dubious way of getting rid of gravity, or up and down, just to reverse the action of it.

But this way is the only way, for we, I have found, cannot conceive it away ; we have to conceive it acting every way, then, affecting each view impartially, it affects none more than another, and is practically eliminated.

The cube had not only to be turned upside down, but also laid on each of its sides and then learnt. There were a considerable number of positions, twenty-four in number, which had to be brought close to the mind, so that the lie of each cube, relative to its neighbours and the whole block, was a matter of immediate apprehension in each of the positions.

If a single cube be taken and moved about, it will be found that there are twenty-four positions in which it can be put by turning it, keeping one point fixed, and letting each turning be a twist of a right angle. The whole block had to be turned into each of these positions and learnt in each.

Thus the block of cubes seemed to be thoroughly known.

At any rate, up and down was cast out. And we can now attach a definite meaning to the expression "casting out the self." One's own particular relation to any object, or group of objects, presents itself to us as qualities affecting those objects — influencing our feeling with regard to them, and making us perceive something in them which is not really there.

Thus up and down is not really in the set of cubes.

Now these qualities or apparent facts of the objects

can be got rid of one at a time. To cast out the self is to get rid of them altogether.

As soon as I had got rid of Up and Down out of the set of cubes I was struck by a curious fact.

If in building up the block of cubes one *goes* to the left instead of to the right, keeping all other directions the same, a new cube is built up having a curious relation to the old cube. It is like the looking-glass image of the old cube. Every cube in the new block corresponds to every cube in the old block, but in the new figure it is as much to the left as before it was to the right. And any set of names in the block so put up gives a shape which is like the shape denoted by the same set of names in the old block, but which cannot be made to coincide with it, however turned about. It is the looking-glass image of the old shape. The one block was just like the other block, except that right was changed into left. Now, was it necessary to cast out right and left as had been done with up and down? or was right and left, as giving distinctions in the block and in shapes formed of cubes, to remain? It seemed as if right and left belonged more to me than to the set of cubes. And yet the right-handed set of cubes could not be made by moving about to coincide with the left-handed set of cubes. And this power of coincidence was the test which had convinced me of the self nature of " Up and Down."

Let Diagram I. represent a small block of cubes. It is itself in the form of a cube, and it contains 27 cubes. For purposes of reference we will give a number to each cube, and the number will denote the cube where it is.

In the front slice are cubes numbered from 1 up to 9, in the second slice are cubes numbered from 10 to 18, and so on. Thus behind 1 is the cube 10. This cube and the cube 11 are hidden, but the cube 12 is shown in the perspective.

Now in this block of cubes there is a part which is known and a part which is unknown. The part which is known is how they come or the arrangement of them. The part that is unknown is the cube itself, repetitions of which in different positions forms the block.

The cube itself is unknown, because, being a piece of matter, it possesses endless qualities, each of which grows more incomprehensible the more we study it. It is also unknown in having in it a multitude of positions which are not known. The cube itself is, amongst other things, a vastly complicated arrangement of particles. Hence, *putting all together*, we are justified in calling the cube the unknown part ; the arrangement, the known part.

The single cube thus is unknown in two ways. It is unknown in respect to the qualities of hardness, density, chemical composition, &c. It is also unknown as a shape. If it really consisted of a certain number of parts, each of which was clear and comprehensible in itself, then we should know it if we grasped in our minds the relationship of all these parts. But there are no definite parts of which a cube can be said to be made up. We can suppose it divided into a number of exactly similar parts, and suppose that all are like one of these parts. But this part itself remains, and the problem remains just the same about this part as about the whole cube.

Now there is a double perplexity: one about the nature of the matter, the other about the cube as to the arrangement of its parts. We will give up any question about the matter of which the cube is composed ; to know anything about that is out of the question. But, supposing it to be of some kind of matter, it presents an inexhaustible number of positions. It can be divided again and again.

Let us look at the block again, and for the moment

dismiss from our minds the question just raised as to the single cubes of which it is built up. Let us look on each of these cubes as a unit. Then two of the units, taken together, form a shape ; three or five of them would form a more complicated shape, and so on.

We can also suppose the cubes away, and think merely of the places which they occupied. In this manner, by first thinking of the 27 cubes, and then simply by keeping the places of them in our minds, we get 27 positions, and in these positions we can suppose placed any small objects we choose. Each of these positions may be called a unit position, and we can form different arrangements of small objects by putting them in different ones of these positions. Now in all this we do not divide the cube up. We simply think of it as a whole—we think of it as a unit. Or if we take the room of the cube instead of the cube, and think of the place it occupies, which I call a position, we do not divide that position up. We take it, if I may use the expression, as a unit position. And *without asking any question as to the nature of these positions, whether they are complicated ideas or not*, we have a kind of. knowledge of the whole block, in that it consists of this collection of 27 cubes, or of this set of 27 positions.

Thus in a rough and ready manner there is something which we can take. If we do not inquire about one of the cubes itself, we are all right ; that being granted we can know the block.

But if we look into what each of these unit cubes, or what each of these unit positions is, we find quite an infinity opening before us. There is nothing definitely of which we can say that the whole unit cube is built up, and each of the positions has a perfectly endless number of positions in it, if we come to examine it closely. All that we can say is that our ignorance

about each of the unit positions is of the same kind
as our ignorance about every other, and, taking one as
granted, we may as well take the 27 as granted; and
so out of a lot of similar ignorances we get a kind of
knowledge of the whole. And this knowledge is not
a mere indefinite thing, but it can be worked at, im-
proved, and made perfect after its kind. For suppose
we limit ourselves to the 27 positions numbered in
Diagram I. Two of these positions form one shape,
three of them will form another shape, and so on. And
in going over each of these arrangements we gradually
get to know the whole set of them which form the
block.

Having given up for the time any question as to the
possible subdivisions of the cube, and looking on each
cube as a unit position, we have 27 positions. These
positions can be taken in different selections, and each
selection is a shape. To know the block or set of
positions means to form a clear idea of every shape, consisting of selections of positions, which can be formed out
of the 27.

But each of the cubes, 27 of which form the whole
block, can be divided up. Each of these cubes contains
a great many positions. There must, for instance, be
positions in each cube for every one of its molecules.

Thus it is evident that the cube supplies an inexhaustible number of positions to be learnt. I call the
cube unknown in the sense that there are a great number
of positions in it which are not clearly realized by the
mind.

By a very simple device it is possible to penetrate a
little into the unknown part. The whole set of cubes
forms a cube. Let us consider the small cube to be a
model of the whole cube. Let us consider it as consisting of 27 parts, each related to the other as the 27 first

cubes were related amongst themselves. Thus the unknown part, the material cube, which is used to build up
the whole, becomes reduced in size. Fig. II. represents
such a cube.

This is the theory. The practical work consisted in
learning the names denoting these smaller cubes in connection with their positions, so that, the names being said,
the small cubes meant were present to the mind, and a
set of names being said, the shape, consisting of a set of
cubes in definite relations to each other, came vividly
before one. A complete knowledge of the block of cubes
would be a complete appreciation of all the possible
shapes which selections of the cubes would form, and
this I strove to attain. Here at length I found real
knowledge, and after a time I was able to reduce the
size of the unknown still further, and to obtain a solid
mass of knowledge fairly well worked all through.

And now it all seemed satisfactory enough. There
was real knowledge in knowledge of the arrangement;
and the material cube, which must be assumed, could be
made smaller and smaller, it could be turned into knowledge, thus affording a prospect of obtaining endless
knowledge. Thus I found the real home of my mind, the
only knowledge I had ever had, and I hoped always to
continue to add to it, and always to reduce the unknown
in size.

Presently, too, the forms of the outward world began
to fall in with this knowledge; and as the mass of known
cubes became larger in number, a group of them would
fairly well represent a wall, a door, a house, a simple
natural object such as a stone or a fruit.

Yet amidst all this delight I became conscious, dimly
enough, of a self-element in the knowledge of blocks.

If, putting up the block of cubes, we go to the left
instead of the right, but in all other respects build up in

the same way, we obtain a block which has a curious relation to the first block.

The ordinary block is shown over again in Diagram III. Diagram IV. is the new block. The new block is like a looking-glass image of the old block. It is just the same, but that left and right is reversed.

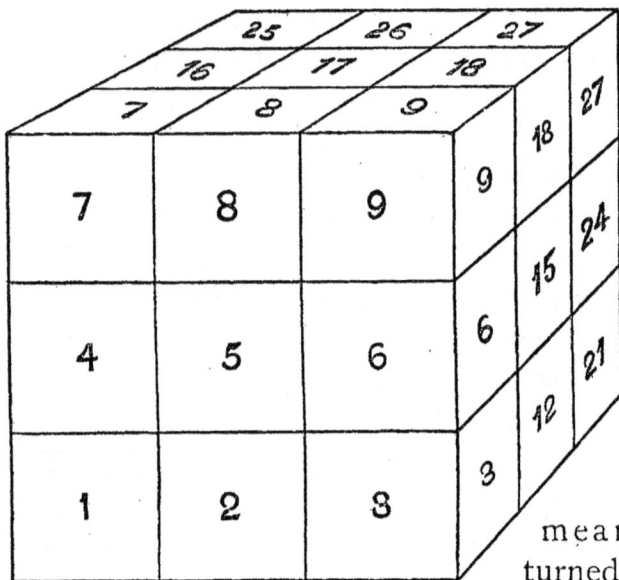

Diagram III. is a block.

Also, if we take selections of blocks we get figures which are just reversed. Thus 1, 4, 7, 8, in Block III., means a figure turned to the right; in Block IV. a figure turned to the left.

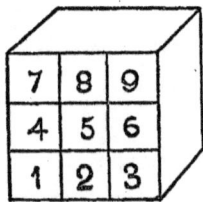

Diagram V.

Again, consider the two figures formed by selecting the cubes 1, 4, 7, 8, 17, from Diagrams III. and IV. respectively. We get two figures which are just like one another as arrangements, but which we cannot turn into one another by twisting.

Considered as arrangements in themselves, these figures and these blocks seem to be identical, for the relationships of cube to cube which are present in the one are all present in the other. But considered as shapes they are not identical. For they will not coincide.

The whole matter becomes much more clear if we consider the relationship between the individual cube used and the block which it forms.

There are two starting-points, either of which we can adopt. We can start with the real material cube, or we can start with the act of arranging. When I speak of the real material cube I do not want to call attention to the kind of matter of which it is composed, or to the nature of matter, but to the fact that it is to be a real cube such as can be made, and which, if one edge or corner be marked, will retain that mark just where is not a product of but an object, with objects in general. the real material the cube shown in is the model on a small scale of the Block III. The numbers in it show the small cubes of which we suppose it to be built up after the pattern of Block III. The numbers also serve to show the distinction of positions— that is, we can refer to the right-hand corner or edge, &c., by saying the numbers of the small cube which lies there.

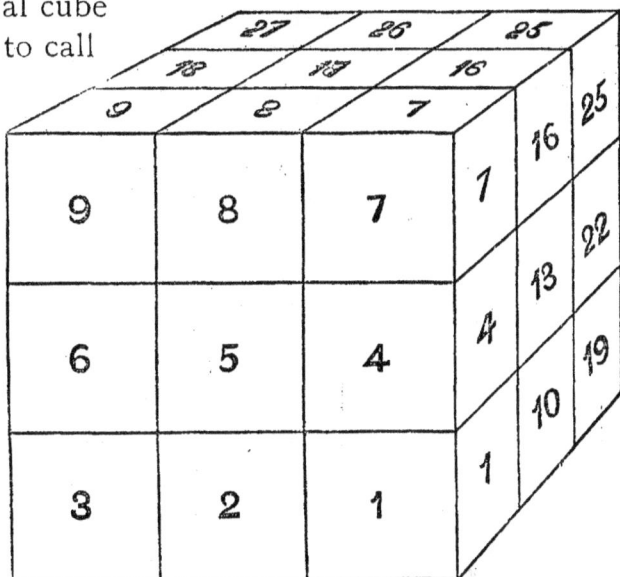

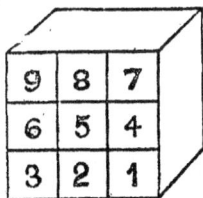

Diagram IV.
is its image block.

Diagram VI.

it is—a cube which the imagination, the properties of Let us start with cube. Let us take Diagram V., which

Now, using the cube of Diagram V. to build up the block in Diagram III. we get a perfectly orderly result, as shown in Diagram VII., and we can go to bigger and bigger blocks, or down to smaller and smaller ones without any hitch. But if we use the cube of Diagram V. to build up the block of Diagram IV., there is a disadjustment which can be discerned in Diagram VIII. Thus, when V. is used to build up III., the small cubes in V. 1, 4, 7, lie in same edge as the cubes 1, 4, 7, in the big Cube III. But when V. is used to build up IV., the small

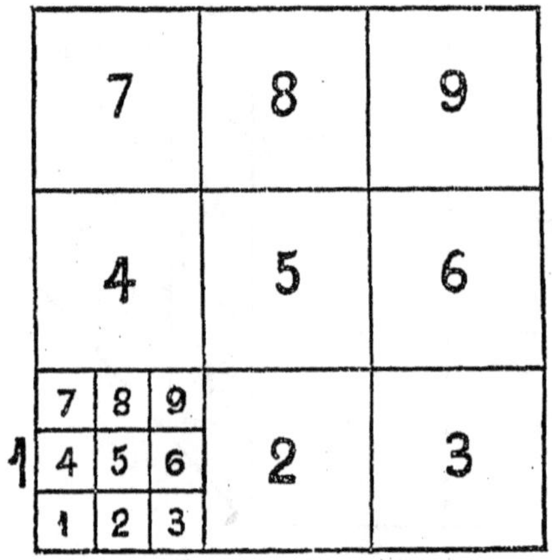

7	8	9	9	8	7
4	5	6	6	5	4

7	8	9	2	3	3	2	7	8	9
4	5	6					4	5	6
1	2	3					1	2	3

Diagram VII.—Block III. built up with Cube V.

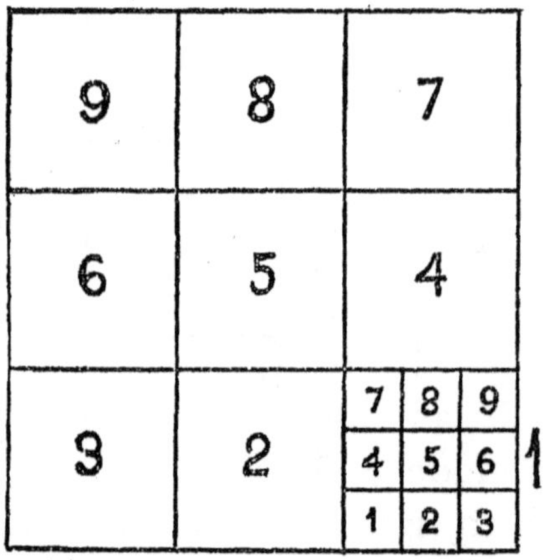

Diagram VIII.—Block IV. built up with Cube V.—a disadjustment.

cubes 3, 6, 9, lie on the edge which is occupied by the cubes 1, 4, 7, in big Cube IV.

Thus, if the same material cube is used, there is a disadjustment, and the figure IV. cannot be considered the same as the figure III. even as an arrangement, for the same parts of the cubes do not lie in an analogous manner. A certain corner of Cube V. is marked with the figure 7; this corner would be on the outside in Block III., but in building up Block IV. it would lie on the inside.

It is somewhat difficult to express this fact, but if the real cubes are looked at it becomes perfectly obvious.

Imagine the whole Block III. to be built up of a number of cubes, every one of which is alike. If the sides of these cubes be distinguished by any markings—if, for instance, the left-hand side is blue and the other sides are each of some special colour, then on building up the whole block the left-hand side of the whole block will be blue.

If, now, the same cubes be taken, and the attempt be made to build up the looking-glass image of the block with them, it will be found that there will be a disadjustment. If the blue sides are made to go to the right, as they must, to form an image block, then some other sides will be in different places to what they should be in order to produce an image of the original block. Although considered as an arrangement of cubes the new block will be an image of the original block, still, looking at the individual cubes of which it is composed, it will be seen that the new block is not an exact image of the old block.

If, however, we take the other starting-point, and, not assuming any fixed fundamental cube, look only at the act of arrangement, the two Blocks, III. and IV., are found to be identical in every internal relationship.

For, taking the act of arrangement as the basis, if, when we have built up the Block IV., we look upon each of the cubes as an arrangement of the same kind as the whole, then the cube 1 in Diag. IV. is represented in Diag. VI. And it is evident that if Diag. IV. is built up out of cubes like Diag. VI., the small cubes, 1, 4, 7, lie in the same edge as the cubes 1, 4, 7, in Diag. IV. Thus it will be found for every relationship in Diag. III. there is an exactly similar relationship in Diag. IV.

In this case if, for the sake of material illustration, we use marked cubes, it seems that we must not suppose each particular cube to have a fixed marking of its own, but that we must suppose the markings to spring up on the sides of the cubes in accordance with the places into which they are put.

There is another manner of regarding the matter which may help to bring out the point at issue.

If we suppose that we are putting up the cubes in one room while another person is putting up cubes in an adjoining room ; if we can tell him what we are doing, using the words right and left, he will be able to put up a block exactly like ours. But if we do not allow ourselves to use the words right and left, but speak to the other person as if he were simply an intelligence without having the same kind of bodily organization as ourselves, we should find that, supposing he could put up the block of cubes, it would be a mere matter of chance whether he had put up the block as we had put it, or whether he had put it up in an image way. And the same with regard to any shape. We could tell him that the cubes should be put together, and we could tell him the relationship which they should have with regard to one another ; but the figure he put up would just as likely be an image of our shape as not.

And we could go on for ever building more and more complicated shapes and telling him to do the same, and no hitch or difficulty would come. But at the end all his shapes might be ours just reversed, as if seen in a mirror.

And if, having put up the block, we coloured the sides of the cube we used as the fundamental cube, and told him how we had coloured it : if he coloured his and brought it to us, and we compared them, his would just as likely be the image of our cube, and not able to be turned into it. So that although, as arrangements, the structures we had put up were alike, still neither of us could use the other's fundamental cube ; and if we ex-

changed the fundamental cubes there would be an in-
consistency in each of our arrangements.

Now, are these blocks of cubes really the same? Are
III. and IV. really the same in themselves, as all relationships in the one are to be found in the other ? If so, the
feeling on my part that they are different, and the inconceivability of their coinciding, must be due to some
self-element which is mixed up with my apprehension
of the cube.

The Block IV. is like the Block III. in its known part—
in its arrangement. It is unlike Block III. in its unknown part—the cube which must ultimately be supposed
as the fundamental cube, by using which over and over
again the whole is built up.

Now, the properties of the unknown part—the little
cube of matter which of some size or another, we must
assume, are so mysterious that one does not feel any
argument very safe which rests on it.

Moreover, there is a very obvious consideration which
reduces the importance of the part played by the
material cube very considerably.

It is possible to consider the Cube V., which is used to
build up III., as the total of 27 cubes.

But each of these cubes—the small cubes in Diag. V.—
can be considered to be made up of 27 still smaller cubes.

By going on in this way we can get our fundamental
cube very small indeed. The difference between the
Cubes III. and IV., in respect to this fundamental cube,
will still remain. But omitting this difference they will
be, considered as arrangements, identical.

To state the matter over again. We start with a real
cube, one inch each way, and build up the block in
Diagram III. with it. If we try to build up the block
in Diagram IV. with this same inch cube, we find that
there is a disadjustment.

But we are not obliged to have our fundamental cube one inch in size. We can take it as small as we like, and build up the block, using a greater number of such cubes. We can take it the twenty-seventh of the twenty-seventh of an inch cube; or, in fact, as small as ever we like. And if we take a very small cube as the fundamental one with which we build up the Block III., then, using this same fundamental cube to build up Block IV., we should find a disadjustment, although this disadjustment would only come in when we come down to the very minute cube, and studied its relationship to the whole Block IV.

Thus, apparently, the Block IV. could never be built up consistently, using as its fundamental cube the fundamental cube out of Block III. But in saying this we have really made an assumption.

It is obvious that Cubes V. and VI., just like Cubes III. and IV., considered as shapes made up of matter, are very different, and could not be shifted one on to the other.

But all our laws and feelings about movements and possibilities are founded on the observation of objects having a certain degree of magnitude.

But the fundamental cube, which we must assume, may be supposed to be of a degree of magnitude less than any known degree.

In cubes of a certain size V. and VI. are different, and cannot be made to coincide.

But we are absolutely unable to say anything about cubes beyond a certain degree of smallness. With cubes of a certain degree of minuteness, V. and VI. might be able to be made coincide.

Thus, for instance, we feel as if we could divide a piece of matter on and on for ever. But chemists tell us that, after a certain number of divisions, the next division would split it up into two different kinds of matter. Since all our reasoning is founded on the behaviour of objects of known size, we can tell nothing at all by inference about the behaviour of very small objects.

It is obvious that, from our customary experience, we can assert absolutely nothing at all about the extremely minute or the extremely large. All reasoning which is founded on the likeness between the extremely small and the ordinary objects of our observation is absolutely valueless as telling us any truth.

Of course, by saying this we have not got rid of the argument for the difference of III. and IV. But we have put the thing from the observation of which that argument is drawn out of the region of known things. We have put it into the hazy land of the extremely minute. Its argument is good, but it depends on its being of a certain size. We suppose it less than that size, and we can consider the subject without regard to its argument.

The question then before me was, Is " Right and Left " to be cast out ? And connected with this was the consideration of whether it was possible for extremely minute cubes to be " pulled through," that is, to be treated somehow which would turn one like V. into one like VI.

Now, if "right and left" was a self-element, it could be cast out ; if it was a permanent distinction in the cubes themselves, it could not be cast out. The thing to do was evidently to try. The method was to learn the cubes over again, in a set of new positions. For every one of the ways in which they were learnt before, there was an inverted or pulled through way to be learnt.

While I was engaged in this attempt another inquiry suddenly coincided with this, and explained it all.

Much has been said about the fourth dimension of

space and the inconceivability of it to us. Now, if there are beings who live in a four-dimensional world, they must feel as habituated to it as we do to ours, and the conceptions which seem so impossible to us must be every-day matters to them. It would be impossible for us to try to enter at once into the serious thoughts of these denizens of higher space. But amongst them there would probably be some with whose occupations we might become familiar, and with whose ideas we might gain some acquaintance. Amongst these beings there must be children, and just as children on the earth gain their familiarity with space by means of bricks and blocks and toys, so these higher children must have their own simple objects wherewith they grow into familiarity with their complex world.

Now it is easy to make a set of simple objects such as these higher children would use. And it seemed a practical thing to do with regard to the conceivability or inconceivability of the fourth dimension to give the matter a fair trial, by going through those processes and those experiences which must be gone through by the beings in higher space to gain their acquaintance with it.

When I say that it is easy to make a set of objects, such as the higher children use, I do not mean to say that they can be made completely in every part at once. But we can make the ends and sides of them, and we can look at the ends and sides of them as they appear to us in space, and we can make up exactly what sides come into space when the simple objects are twisted and moved.

Just as a being living on a plane could tell about all the faces and edges of a cube or other simple solid figure by looking at what he could see when the cube was laid on his plane, and when it was twisted and laid down again;

so we can tell all about the sides, faces, and edges of a higher solid.

And the project seems less uninviting if we reflect on how complicated a matter the formation of our own conceptions of a solid are. What a lot of faces and edges a cube has! And, moreover, it must be remembered that we never touch or see a solid; we only see the surface and touch the surface. If we cut away the surface that we first saw or touched, we come on another surface, and so on.

Now, of course, the surfaces of a solid are given to us by nature in their right connection and relation. Each of the edges of the cube, for instance, can be noticed and remarked without any difficulty, and they are all on the same bit of space, to be looked at one at the same time as another.

But the sides, faces, and edges of a higher solid cannot be in our space all at once. They must come separately, be looked at one by one.

Thus a being in a plane could not see the lower side and the front of a cube at once. He would first have to look at the lower side as the cube rested on his plane, then if the cube were turned over he would see the front, and the lower side would be gone. If he got the set of right appearances which a cube would present to him when, turning about in a systematic way, it came at intervals into his plane, and if, moreover, he fixed his mind on these appearances, he might at last, if it was in him, rise to the conception of a cube as we know it.

Now, the parts by which a higher solid comes into our space are solids, and what we have to form is a set of solids coming and going in a systematic way, as the higher figure is moved about in a systematic way.

This afforded a welcome exercise, for conceiving the

solid shapes, and how they went and came, increased my familiarity with the set of cubes.

Moreover, in trying to get the piece of ignorance—the necessary real cube—as small as possible, I had got the block which I knew to a somewhat fine state of division, and could, by picking out a particular set of cubes from the whole number, obtain a mental model of any shape I wanted. The whole block of cubes formed a kind of solid paper in which one could mentally put down any solid shape one wanted. And just as it is a great convenience to have a piece of paper for drawing figures one wants to think about, so it was a great convenience to have this solid paper.

The subject, however, abounds in abysses for stupidity to fall into, and I had to clamber out of each of them ; so it took me several years before I got quite on the right tack. Then it was easy enough : any one in a few weeks could learn to conceive four-dimensional figures. Not only is it easy, but there are abundant traces that we do it continually without being aware of it. I am sure if the loveliness of the work while one is doing it, and the simplicity and self-evident nature of the results when obtained, were generally known, it would be a favourite amusement.

Now one of the first things that presented itself to my attention when I began to move the four-dimensional figures about was a fact which bore curious reference to my difficulty about the fundamental cube. If the reader remembers, it seemed to me as if the cube out of which the whole block of known cubes was built ought to be able to be inverted. That is to say, it seemed to me that there was a self-element present in my knowledge of the cubes. But in order to cast out that self-element the fundamental cube which lay at the basis of the whole block would have to be able to be inverted, or pulled through.

Now I found that when I took a four-dimensional figure which came into space by a cube—that is, a figure which rested on space by a cube, or one of whose sides was a cube—when I took a figure of that sort up in the fourth dimension and twisted it round and brought it down again, this cube would sometimes be inverted or pulled through—although I had done nothing to it, but had simply twisted the whole figure round without disturbing the arrangement of its parts.

Thus evidently to a higher child it would be no more difficult to invert or pull through a cube or a figure than it would be to me to twist one round.

Hence it was obvious that right and left was really a self-element in my block of cubes. I being in our space was under a certain limitation, and that limitation made me feel as if a right-handed arrangement was different from a left-handed arrangement.

A being who was not limited as I was would see that they were one and the same. Hence, in knowing the set of blocks it was necessary to cast out " right and left," and the names had to be learnt over again in new positions.

Thus it is evident that there are three expressions which may be considered in reference to a knowledge of a block of cubes as almost identical : " Casting out the self "—" Seeing as a higher child "—and thirdly, " Acquiring an intuitive knowledge of four-dimensional space."

Thus, taking the simplest and most obvious facts—the arrangement of a few cubes—we found that there was a known part and an unknown part; the known part corresponding to our act of putting, the unknown part the cube which, of some size or another, must be taken as given in the external world. Then there was obviously a self-element present in the Up and Down felt as in the cubes. This being removed, Right and Left had also to go. So, to get the knowledge of this simple set of objects clear of self-elements, two universe transforming thoughts have to be used ; and when these thoughts are thus in-corporated the cubes become different.

It will be obvious to the reader that in these pages I have merely touched the surface of the subject. But the deeper matters which are contained in the knowledge of a block of cubes are difficult to express, and are so mixed up with the practical work, as far as I conceive them at present, that it is best to consider in some detail the applications to the world about us of those truths of which we have already got a clear apprehension from the block of cubes.

Instead, then, of going on, let us conclude the present paper by going back, and taking a simple instance of the general truth that progress in the knowledge of a block of cubes is casting out the self.

Let the reader turn to Diagram I. and make out the shape which the following numbers denote—namely, 1, 4, 5. If the following numbers be said, 18, 27, 26, it will be found that they denote the same shape, but in a different position. Now if the block of cubes be well known, these two sets of names, 1, 4, 5, and 18, 27, 26, ought to convey instantly to the mind the same idea. However quickly they are realized, it ought to be evident that they are the same shape.

And a good deal of the practical work in learning a block of cubes consists in gaining this faculty of immediate apprehension. But when it is gained it is seen to consist much more of getting rid of an imperfection than in being any real advance. For if the two shapes are identical we need not ask ourselves how it is we see them as the same, but we have to ask ourselves what is the reason why we do not recognize their identity ; and the answer evidently is that, if we do not recognize their identity, it is due to the particular relationship of each shape to ourselves. One is down on our left hand, another is up on our right, and they are turned relatively to us different ways. Now these differences, which are merely relative to us, we impress upon the shapes, and really feel the shapes to be different. The practice consists in getting rid of the influence of these self-elements, so that two shapes, however complicated, being alike, when their names are said, we feel them to be alike without calculation or reflection. Thus the power of seeing likeness and analogy in this domain is merely another name for the power of casting out the self-elements from our mental presentation of any objects with which we come into contact.

www.ingramcontent.com/pod-product-compliance
Lightning Source LLC
Chambersburg PA
CBHW051344200326
41521CB00014B/2474